SpringerBriefs in Business

AF212621

More information about this series at http://www.springer.com/series/8860

Nobuyuki Tokoro

The Smart City and the Co-creation of Value

A Source of New Competitiveness in a Low-Carbon Society

Springer

Nobuyuki Tokoro
College of Commerce and Graduate School
of Business Administration
Nihon University
Setagaya-ku, Tokyo
Japan

ISSN 2191-5482 ISSN 2191-5490 (electronic)
SpringerBriefs in Business
ISBN 978-4-431-55844-6 ISBN 978-4-431-55846-0 (eBook)
DOI 10.1007/978-4-431-55846-0

Library of Congress Control Number: 2015954614

Springer Tokyo Heidelberg New York Dordrecht London
© Springer Japan 2016
This work is subject to copyright. All rights are reserved by the Publisher, whether the whole or part of the material is concerned, specifically the rights of translation, reprinting, reuse of illustrations, recitation, broadcasting, reproduction on microfilms or in any other physical way, and transmission or information storage and retrieval, electronic adaptation, computer software, or by similar or dissimilar methodology now known or hereafter developed.
The use of general descriptive names, registered names, trademarks, service marks, etc. in this publication does not imply, even in the absence of a specific statement, that such names are exempt from the relevant protective laws and regulations and therefore free for general use.
The publisher, the authors and the editors are safe to assume that the advice and information in this book are believed to be true and accurate at the date of publication. Neither the publisher nor the authors or the editors give a warranty, express or implied, with respect to the material contained herein or for any errors or omissions that may have been made.

Printed on acid-free paper

Springer Japan KK is part of Springer Science+Business Media (www.springer.com)

Preface

Is it possible to establish a sustainable society in which human economic activities coexist in harmony with the maintenance of the global environment? This is a topic that has attracted significant interest in the fields of natural science and social science in recent years. For sometime now, there have also been indications that the massive expansion of human economic activities is exceeding the limits that nature can tolerate. *Limits to Growth*, a report published in 1971 by the think-tank Club of Rome, shocked the world at the time by presenting a catastrophic scenario of a global environment in distress due to severe depletion of resources, widespread pollution, and overpopulation. Nearly half a century has already passed since then, and the predictions of the Club of Rome seem to be vindicated as the world indeed seems to be heading in the direction of the scenario it depicted. The world population has topped 7 billion, and the problem of the depletion of oil resources is becoming a real possibility. Initiatives to reduce carbon dioxide emissions that are the cause of global warming have made little headway, and emissions of the world as a whole continue to rise. Have our problems reached a state where we no longer have the means to solve them?

The answer to that question is that not everything on the horizon is bleak, and there are definite signs that give us hope for the future. One of these is the creation of innovation by the corporate world. Although in recent times more than a few companies have been severely chastised and named by environmental conservation groups as culprits of environmental pollution, at present innovations forthcoming from the corporate world are beginning to be viewed as the only means for resolving environmental problems. For example, car makers are focusing on the development of electric cars and cars powered by fuel cells to replace existing gasoline-powered cars. If these cars become widely used by general consumers, they will contribute considerably to the reduction of CO_2 emissions. The same can be said of the widespread generation of electricity using renewable energy such as photovoltaic power generation and wind power generation promoted by manufacturers of electrical appliances and equipment. The development of new technologies and products by such companies has the power to significantly change today's society. A look at past changes in society makes this point clear.

This book examines case studies of smart cities that are currently attracting attention as promising initiatives in the establishment of a low-carbon society and attempts to verify the achievement of companies in these initiatives from the perspective of establishing competitive advantages. Smart cities are experiments in making entire communities "smart" by linking various products and services created to conserve the environment and represent an overall aggregate of innovative products and services achieved by various companies. Moreover, companies that are the main agents in the creation of smart cities are not individual companies of a single industrial sector but comprise companies hailing from diverse sectors. In other words, innovations created in a smart city setting are products of co-creation by companies of disparate sectors.

The case studies of smart cities taken up in this book are all projects that began in Japan around 2010. Participating in the three projects examined are major corporations representative of Japan such as Panasonic, Toshiba, Hitachi, and Nissan Motor, and all three projects are different in nature. The purpose of this book is to carefully examine the kind of competitive advantage the respective companies achieve through their participation in these projects. To do this, the book establishes a theoretical framework known as a "relationship-based strategy" and confirms the processes of co-creation among companies and the establishment of competitive advantage.

Existing theories for analyzing the competitive advantages of companies are all based on competition among companies. Essentially, the underlying theory is established on the basis of how one company can defeat its competitors in the market. Even in the so-called positioning view and resource-based view, which are representative existing theories concerning competitive advantage, despite differences in the focus of attention regarding the source of competitive advantage, there is a common understanding that a competitive environment with other companies creates competitive advantage. This book, on the other hand, adopts the view that it is not competition with other companies but co-creation with other companies that creates competitive advantage. During the process of co-creation with other companies, the technologies, know-how, and knowledge that companies have are exchanged and integrated, thereby creating new value. This book tells the story of how that value becomes a competitive advantage for companies. In this story, the smart city projects can be considered the stage where various diverse companies co-create and actually bring about new value.

In bringing this book to publication, as the author I have received the kindnesses of many people along the way. In particular, my colleague Professor Mitsuru Kodama, who is also faculty of Nihon University, has provided me with valuable advice in a wide range of areas including the establishment of theories, drawing of implications, and question items for the field survey. Being blessed with the opportunity of an intellectually stimulating workplace, environment afforded me the rare experience of writing this book. I would like to take this opportunity to express my heartfelt thanks to Professor Kodama. Moreover, during my field research, I had the opportunity to speak with many corporate stakeholders including dSPACE Japan, Panasonic, Toshiba, and Hitachi, who provided me with valuable information. I would like to express my deep appreciation to them too. To publish this book, I was fortunate enough to receive a grant from my employer, Nihon University, and I would like to express my gratitude to the university for providing me on a daily basis with a richly rewarding research environment. In closing, I would also like to take this opportunity to express my gratitude to Springer for providing me with the opportunity to publish this book and to all of those in the editing department for their support.

Looking forward to a better future.

Nobuyuki Tokoro

Contents

About the Author

Nobuyuki Tokoro is a Professor of Business Administration at the College of Commerce and the Graduate School of Business Administration at Nihon University. He specializes in environmental management and innovation and is particularly interested in the relationship between corporate technical innovation and competitiveness in low-carbon societies. He has authored and published many books and papers both in Japan and internationally on the subjects of environmental management, innovation, and competitiveness. His 2005 book *Environmental Management* (Shinka suru Kankyo Keiei) published in Japan was awarded an association award from the Sustainable Management Forum of Japan.

Chapter 1
Smart Cities and Competitive Advantage: A New Perspective on Competitive Edge

1.1 Introduction

Modern society faces a wide range of issues, and one of the greatest that human beings have to solve in the 21st century is global warming. In order to limit temperature rises, we have to dramatically cut emissions of carbon dioxide, which is one of the causes of global warming, but the close linkage of this problem with issues of economic development means that various interests are involved, and it is common knowledge that the world as a whole is making less headway than expected.

In these circumstances, the construction of smart cities is attracting attention. Smart cities use information and communications technology to optimize electricity, water, telecommunications, transport systems, and other social infrastructure that form the foundations of urban activity and create eco-friendly urban areas that aim to reduce energy consumption and carbon dioxide emissions. We can anticipate that building such cities in various places around the world will encourage the development of a low-carbon society in general and play a role in ending global warming.

The purpose of this chapter is to observe the building of smart cities from the viewpoint of corporate competitive advantage and search for a source of new competitive advantage in the process. Smart cities can be analyzed from various angles and approaches including urban planning, urban administration, information technology, networks, urban civilization, and innovation. This book takes its approach from the corporate competitive standpoint that companies are the main players in smart city construction, and that the corporate acquisition of a new competitive advantage through these construction projects further encourages the expansion of smart cities, and leads to the realization of a low-carbon society.

This chapter begins by outlining the current state of smart-city building projects proceeding across the world and reviewing prior research into smart cities. It then considers existing theories of corporate competitive advantage, and aims to discuss concepts focusing on a new theory of competitive advantage presented in the text.

© Springer Japan 2016

N. Tokoro, *The Smart City and the Co-creation of Value*,
SpringerBriefs in Business, DOI 10.1007/978-4-431-55846-0_1

1.2 Progress of Smart City Building Projects Around the World

Interest is growing in smart city construction as a solution to the increasingly serious problem of global warming. Smart cities aim to use information and communications technologies to promote energy efficiency and create low-carbon city planning with reduced carbon dioxide emissions. But the advantages of smart cities don't stop there. Anticipation is growing that smart cities can be a powerful trump card to solve the various problems faced by cities today including urban population growth, aging society, traffic congestion, and security.

Smart City construction projects are making progress throughout the world, with numerous projects ongoing both in industrialized and developing countries. Here, I will give an outline of the current state of smart city construction projects implemented in different parts of the world while considering the Nikkei BP Cleantech Institute research titled "World Smart City General Survey 2012" and "Smart City Report 2013" (Nikkei BP Cleantech Insitute and Techno Associates 2011, 2012).[1] According to the institute's surveys, 608 smart city construction projects were implemented around the world in 2013. Broken down by country and region, China accounts for the lions' share with 225 projects, followed by North America with 124, Europe with 91, Asia excluding China and Japan with 78, Japan with 63, Africa with 17, and South America with 10 (Table 1.1).

The contents of the 608 projects are diverse, but those with elements of urban development account for up to 315, making around half of the total. Classified according to developed or developing nations, the latter account for 232, or around two-thirds of the total. Among these, projects involving urban development in China account for 143, by far the greatest number. China is urbanizing at breakneck speed. Twelve million people are migrating from rural areas to the city each year, with the result that Chinese cities face a wide range of problems including population growth, air pollution, and traffic jams, and China is undertaking urban development against that backdrop. Outside of China in Southeast Asia, the Middle East, and African countries the concentration of people in large cities has become an issue, and new urban development projects are starting up one after another.

Urban development projects in these developing countries often take the form of new urban development happening alongside industrial zone development. In many cases this takes the form of developing large-scale new cities that combine the creation of new residential neighborhoods on vacant plots or reclaimed land with industrial zones having fully equipped infrastructure in order to lure industry. These developments are driven by the aim to guarantee employment for an influx of residents by encouraging industry. For example, in China a project (Tianjin ecological region) is underway to build a new city to hold around 400,000 people in the saltpan of Bohai Gulf. This project aims to combine the construction of residential

[1]The data relating to the smart city projects described above all appear in Nikkei BP Cleantech Institute's "World Smart City Survey 2012".

Table 1.1 Each of the 26 target keywords in 608 projects

Category	Target no.	Category	Target no.
Smart grid	188	Water treatment	44
Introducing recyclable energy	165	Cogeneration	37
Urban development (redevelopment)	142	Aging society initiatives	35
Technology demonstration	132	Health and welfare	32
IT development	125	Smart buildings	30
Urban development (new)	100	Storage	30
Engineering zone development	74	Restoration	16
Administrative services	73	Smart villages	13
Service demonstration	71	Car sharing	13
Introducing EV	62	Hydrogen supply	12
Next-generation transport system	57	Ancillary services	4
Smart house	52	Smart factory	3
Environment protection	45	Marketing	2

Source Smart City Report 2013, Nikkei BP, p. 10

areas with incentives to attract industries involved with environment, energy, research and development, finance, and outsourcing services. Meanwhile, in India, a development plan covering a wide area from Delhi to Mumbai, known as the Delhi-Mumbai Industrial Corridor, is underway. The project involves building dedicated freight railroads over the 1500 km stretch between Delhi and Mumbai, and creating residential districts incorporating housing and commercial areas within a 150 km periphery on either side of the tracks together with industrial zones, logistics bases, power plants, roads, harbors, and other infrastructure planning.

Set against this, provision for industrial zones in developed countries has completed the first stage, and a feature of these projects is that many aim to redevelop existing cities in order to resolve the various problems faced by maturing cities of the emission of carbon dioxide, traffic congestion, and a worsening living environment. Smart grid projects are often implemented in the developed countries of North America, Europe, and Japan. In the case of the United States, its power networks are becoming increasingly decrepit, and a national strategy is being rolled out to solve this problem with the introduction of smart grids. The American Recovery and Reinvestment Act (ARRA) passed by President Barack Obama in February 2009 provided financial assistance to establish a project around a smart grid core.

In Europe and Japan, circumstances are somewhat different. Projects are strongly oriented toward introducing smart grids to create renewable energy and stabilize unreliable electric power networks. In Europe, especially, the European Union has set a renewable energy target of 20 %, and is developing smart grid energy projects to realize it. In Japan's case, the earthquake that struck the northeast of Japan on March 11, 2011, has had a huge impact. After implementing such measures as planned power shutdowns and a 15 % reduction in energy use, power supplies and energy plans are under review, and these plans incorporate the installation of smart grids.

Although many of the projects to introduce renewable energy and demonstrate technologies are undertaken in developed countries, developing countries are also undertaking a number of large-scale projects that merit attention. For example, the Masdar City project carried out in the desert on the outskirts of Abu Dhabi in the United Arab Emirates will involve the construction of an artificial city to accommodate a population of around 50,000 people, and will have its energy needs supplied solely by renewable energies such as solar power.[2]

Meanwhile, although projects to cope with the aging of society, health, and welfare are small in number, they hold great potential. These projects are mostly concentrated in industrialized countries. In the case of projects to cope with the aging society, the total number of 35 breaks down as 16 in Japan, 11 in Europe, and 8 in North America. In the future the demand for all three types of projects is anticipated to rise, including in developing countries, which will have the potential to capitalize on the experiences of developed countries.

1.3 Review of Previous Research

As already mentioned, research into smart cities is currently being carried out vigorously against a background of growing social concern, and is taking a range of analytical approaches. Here I will look at typical examples of previous research and undertake a general review. The most common approach takes the perspective of urban planning. For example, studies undertaken by 36 researchers (Rassia and Pardalos 2014) on an international scale are leading to wide-ranging findings in architecture, engineering, and related areas. Research into sustainable technologies and potential energy to make cities smarter as well as research on future urban energy systems is being discussed and the smart city analyzed, mostly from the viewpoint of urban engineering. In addition, interdisciplinary, international joint research by 37 researchers (Ercoskun 2012) investigated the technological and social problems facing the building of smart urban planning and design. This book focuses on the concept of "resilience," and discusses resilience in various settings using eco-technologies.

Meanwhile, within the urban planning approach, some research puts an emphasis on urban administration. For example, Herrschel (2013) analyzes smart city construction from the viewpoint of regionalism. He focuses on the policy-making mechanism in the process of making cities smart, and cites the case studies of Vancouver and Seattle while considering the impact of the regions' special circumstances in the process of becoming smart. Similarly, the research of Gibbs et al. (2013) analyzes smart-city construction from the viewpoint of urban administration.

[2]The Masdal City project is a model for smart city construction projects in emerging countries, and has generated great interest in developed countries. Participants include Siemens of Germany, General Electric of the U.S., and Mitsubishi Heavy Industries of Japan.

They emphasize that urban planning for smart city construction should be based on the three visions of economic growth, linking to the ecosystem, and social equality, and argue that urban spatial development influences existing urban social strata, political culture, and economic base. Also, Shaw (2013) takes up the case of the Melbourne Docklands redevelopment, undertakes analysis with a project spanning 20 years through diverse stories, and gives an affirmative evaluation of the project's sustainable development. In contrast, Tretter (2013) cites a case study in Austin, Texas while taking the negative view that smart cities cannot solve many of the problems facing urban areas, including homeless aspects but eco-friendly.

Townsend (2013) takes a historical viewpoint in considering capabilities in the areas of urban planning and design and information technology that the industrial city has developed from its sudden rise in the 19-century to the present day. He also analyzes how the two global trends of rapid urbanization and the spread of ubiquity collide, and how technology will impact the city of the future. His analytical approach incorporates the perspective of the theory of urban civilization. Meanwhile Deakin (2014) approaches the smart city theme from the viewpoint of innovation and competitiveness. This joint study from 12 researchers focuses on the governance and modelling processes by which cities migrate from intelligent to smart. The study emphasizes the building of urban innovation networks and creative partnerships as well as the development of learning and knowledge transfer and skills development that marks out truly smart cities as distinct from digital and intelligent cities.

Also, Campbell (2012) assesses the smart city from the viewpoint of urban learning, and considers the relationship between the urban learning process, innovation, and competitiveness. Campbell looks at how creative, continuous learning and innovation is necessary to build truly smart cities. He incorporates case studies of specific cities such as Amman, Barcelona, Portland, and Seattle to show how urban networks function and analyzes the mechanisms by which breakthroughs in learning and innovation occur.

1.4 Examining the Theoretical Framework of Competitive Advantage

So while it is clear that the smart city's analytical framework is broad and varied, this book takes a further approach from the standpoint of corporate competitive advantage. As touched on in the review of previous research, Deakin (ed.), Campbell, and others have already carried out analytical research on smart cities from the viewpoint of innovation and competitiveness, but while this research discussed the issues of innovation and competitiveness in relation to urban networks and the learning process, companies were not the target of analysis. In contrast, this book moves away from existing research by focusing on discovering new sources of competitive advantage for diverse companies involved in the process of cooperating to build smart cities. With this analytical approach, it is essential to organize the theoretical framework regarding issues related to corporate competitive advantage from the beginning. Various theoretical frameworks to analyze corporate competitive advantage already exist. Here, taking up two characteristic ways of thinking, I would like to deepen understanding regarding the problem of what a competitive advantage means. This will be covered in the next chapter, which will take the discussion further.

1.4.1 The Positioning View

The Positioning View, an idea put forward by Porter (1980, 1985), is a model theoretical framework for analyzing corporate competitive advantage. The features of the positioning view are a focus on the external environment surrounding companies, especially structural barriers, and the idea that choosing markets with desirable structural barriers from a profits viewpoint can lead to a competitive advantage. This idea is based on an area of economics that looks at the perspective of industrial organization. Industrial organization theory suggests that high industry profitability due to structural barriers should be eliminated, whereas the positioning view takes the opposite perspective that maintaining structural barriers is desirable from the viewpoint of corporate profits, and in these environments companies find it possible to gain a competitive advantage by positioning themselves in the market.

The positioning view cites five "forces" as structural barriers in the external environment: fierce intra-industry competition, the threat of new entrants, the threat of alternative products and services, suppliers' bargaining power, and buyer's bargaining power. In short, the more these five factors are present, the lower the profitability of firms in the industry. When structural barriers are low, it becomes difficult for companies to gain a competitive edge.

To begin with, fierce competition within an industry relates to the problem of the number of companies it contains. If an industry contains just one company or else a very small number of companies, competition is moderate and the profits are taken by a limited number of enterprises. Considered from the viewpoint of the theory of industrial organization, such a situation, rather than being desirable, should be eliminated; however, the positioning view holds that choosing such markets can lead to high profits. Conversely, if large numbers of companies jostle in an industry, competition intensifies and each company enters into a scramble for profits. Selecting such markets cannot be considered a wise choice.

Next, the threat of new entrants is generally considered to be an issue of entry barriers, which are typically regulated by the government. Governments create regulations for companies hoping to enter the market in order to protect specific industries. Such regulations are frequently seen to inhibit the entry of new players, but for companies well positioned in the industry the regulations act as a bulwark and create an attractive enterprise environment.

What about the threat of alternative goods and services? These are goods and services that satisfy similar customer needs in a different way. For example, kerosene, gas, or electric stoves use different energy sources to keep houses warm in winter but match the same customer needs for home heating. One point to note about the threat of alternative products and services is that they pose a potential threat in cases beyond the simple one mentioned. Let's consider cell phones. Many people would come up with landlines or public phones as alternative products, but potentially there are others. Cell phones carry watches and cameras as secondary functions, so the spread of cell phones could impact the watch and camera industries. If large numbers of people use cell phones to access the internet, the PC, TV, and newspaper industries could also be affected. In this way, markets where the threat of alternative products and services is high cannot be said to be a desirable environment from a corporate viewpoint (Aoshima and Kato 2003).

The bargaining power of suppliers and buyers relates to the issue of profit-sharing between both sides, with bargaining power generally determined by the relative strength of the parties. In such cases, the power relationship depends on which side has the right to determine the price, with the side possessing this right having the greatest share of profits. For example, with buyer bargaining power, the issue is that the power tends to concentrate on the buyer (customer) side. When products and services have a high buyer concentration, the buyer has strong bargaining power and holds the right to determine prices. Conversely, when they have a low buyer concentration, the companies concerned can easily find alternative customers, and this increases the bargaining power of the companies concerned. A similar logic holds for the bargaining power of suppliers.

In this way, the positioning view considers a company's external environment, especially the existence of structural barriers protecting corporate profits, and holds that differences in the market environments where companies are positioned influence corporate profits and competitiveness. However, this view cannot explain the difference in competitiveness among companies positioning themselves in the same market environment. On this point, the positioning view is widely known to have reached its theoretical limit.

1.4.2 The Resource-Based View

The resource-based view got its start as an attempt to overcome the theoretical limitations faced by the positioning view. That is, the positioning view, which seeks structural causes in the external environment for corporate competitive advantage, is unable to explain the factors behind competitive advantage among companies positioned in the same external environment. In this context, a new way of thinking (Wernerfelt 1984, 1995; Rumelt 1984, 1991; Barney 1986, 1996) appeared that focused on a company's internal resources and capabilities rather than the structural factors of its external environment, and held that a company's competitive advantage was linked to these differences. The resource-based view focuses on a company's internal resources and holds that competitiveness arises from its accumulated resources, knowledge, and capabilities.

This new thinking regarding competitive advantage arose against the background of breakthroughs by Japanese companies in the 1980s. At that time, Japanese firms in automotive, electronics, and other markets acquired an overwhelming competitive advantage, but the consensus of management scholars' analysis attributed this to outstanding resources and capabilities accumulated within the company rather than to external structural factors, and interest grew in the notion of internal resources and capabilities as a source of competitive advantage. The concept of core competence also arose from observation of Japanese business (Prahalad and Hamel 1990; Hamel and Prahalad 1994). So just what are these internal resources that become a source of competitive advantage? To answer this question, I will describe the features of the resource-based view below.

Scarcity

We know that some companies hold an advantage over competitors in a market, and the reason is that they possess specific resources in-house. Naturally, competitors will try to acquire those resources for themselves. However, those resources are rare, and competitors will find it difficult to gain a competitive advantage in markets when they cannot easily be acquired. For example, if a company's engineers have developed advanced black-box technology that has become a source of competitive advantage in products, competitors will be pressed to choose between developing similar technology of their own or headhunting their rival's engineers. However, in cases where it is not easy to acquire the talent capable of developing the advanced technologies created by those engineers, or where the likelihood of the original company's engineers being poached through mismanagement is low, the competitors will be unable to obtain the advanced technologies that form the source of competitive advantage. The engineers become a scarce resource positioned as a core competence of the original company.

The Difficulty of Imitation

If resources that cannot be easily copied by competitors accumulate within a company, its competitive advantage becomes sustainable. This raises the question of what conditions can be put in place so that competitors cannot easily copy these resources. One that comes to mind is the cost of copying. Knowing the source of competitive advantage that others want to imitate or the means by which copying can be accomplished leads to situations that involve copying costs. If competitors are unable to obtain another company's resources without paying a price higher than that paid by the company possessing the resources, that cost gap maintains the source of competitive advantage.

A second condition might be a case where copying is methodologically problematic. In this situation, competitors would not know how to imitate even if they were prepared to spend large sums. For example, Toyota Motors' competitive advantage is widely known to lie in its “kanban” or “just-in-time” system of production, and it is said that competitors tried but failed to create similarly efficient systems.[3] The reason is that the Kanban method is not simply superficial production expertise; its essence is said to lie in the relentless cost-cutting efforts emanating from Toyota’s corporate culture. This hidden competitive resource is difficult to discern from the outside and extremely difficult to copy.[4]

Consistency with Customer Values

Possessing scarce resources within a company does not inevitably lead to a source of competitive advantage, even with models that are difficult to copy. For example, even if a company has highly advanced product processing technology that rival companies find difficult to copy, that technology has no meaning unless it confers a competitive edge on products. Japan’s leading electronics manufacturers currently find themselves in just such a position. Sony, Panasonic, and Sharp possess outstanding technologies in the field of high-resolution TV that South Korean and other rivals find difficult to copy, but these technologies do not necessarily constitute a competitive advantage for the reason that the technology is misaligned with customer needs. Many customers are not looking for high resolution TVs with advanced technologies. As long as a certain standard of quality is guaranteed, most customers are prioritizing price. Thus what is important is consistency with customer values. Possessing scarce resources, even if difficult to imitate, cannot confer a competitive advantage unless these resources are consistent with customer value.

This resource-based view focuses on internal corporate resources and considers the source of competitive advantage as combined resources from the conditions of scarcity, copying difficulty, customer value, and consistency. From this perspective, the foci of the resource-based and positioning views differ, but the two concepts are not incompatible, nor do they involve a trade-off. I will go into these points later.

1.5 Co-creation and Competitive Advantage

The theoretical frameworks on the competitive advantages of companies are generally divided into the positioning view focused on a company’s external environment and the resource-based view focused on its internal resources. As seen above, these two viewpoints contrast with each other and can be considered incompatible.

---

[3]For example, in the 1980s General Motors established a joint venture with Toyota known as New United Motor Manufacturing, Inc. (NUMMI). GM hoped to study Toyota’s Kanban system but was unable to copy it.

[4]Hiroyuki Itami emphasizes the importance of invisibility as a source of competitive advantage. Itami and Karube (2004). Strategy and Logic of Invisible Assets. Tokyo: Nihon Keizai Shinbun-sha.

When analyzing corporate competitive advantage in the academic world, scholarly disputes frequently arise over whether to take the positioning or resource-based view. Nevertheless, these two approaches actually complement each other, with each one limited to viewing one aspect of corporate competitive advantage. Put another way, if a company has a competitive advantage in the market, that advantage should be analyzed both from that company's internal and external factors.

For example, Toyota Motor Corporation is one of the world's most competitive auto manufacturers, and its advantage is often analyzed from the theoretical framework of the resource-based view. This interest focuses on the Kanban and "just-in-time" methods embodied in the Toyota Production System (TPS), and these internal resources are analyzed as the source of Toyota's competitive advantage. Indeed, TPS is a competitive advantage that rival companies cannot easily copy, but external environmental factors are greatly involved in maintaining this TPS. That is, Toyota's ability to keep the just-in-time production system depends on a strong relationship with component manufacturers. Without their cooperation, the system could not be maintained. According to the positioning view, the system is achievable because Toyota has strong bargaining power vis-à-vis its component suppliers. Clearly, it is necessary to analyze the TPS, which is often cited as Toyota's source of competitive advantage, from the dual perspective of the positioning and resource-based views.

It is important to analyze a company's competitive advantage from a range of perspectives. Factors that confer a competitive advantage exist both in external and internal environments, and competitive advantage is created from the complex interaction of these factors. This also holds true for the competitive advantages of smart cities that this book is concerned with. When companies gain a competitive advantage through the building of smart cities, the analytical perspective needs to be multifaceted. Confirming this recognition, the analytical perspective of smart cities and corporate competitiveness in this book differs from that of traditional competitive edge theories.

1.5.1 Building a Competitive Advantage Through Collaboration

The existing positioning and resource-based theories of competitive advantage assume competition among companies. That is, to succeed in competition among companies requires a profitable external market structure for a company and the intention to possess internal resources that competitors cannot imitate easily. In this context, competitive advantage is considered to derive from competition among companies. In contrast to this approach, this book presents a new focus on competitive advantage that emphasizes co-creation over corporate competition. That is to say, the concept's basic premise considers the principles of competition from the standpoint of co-creating new value by collaborating with other companies rather than requiring the defeat of weaker rivals. This book analyses smart cities and corporate competitive advantage taking into account the qualities of smart cities, and considers that competitive advantage can be thought to arise from the building of co-creation rather than competition among companies.

The building of smart cities requires collaboration among a wide variety of companies involved in such areas as electric power, automobiles, gas, water, housing, and finance. To give an example of the smart houses that comprise the smart cities, the electricity generated by solar panels installed on the roof is sent to the car in the parking space, where it is stored in the vehicle's lithium ion battery. At night and during emergencies this power can be used and integrated with the household's internal network to provide energy "visibility." Moreover, financial services can be accessed from the living room. In such ways, by freely exploiting information and communications technology, residents can enjoy a hitherto unimaginable level of convenience. These high-value added features of the smart house are not produced by competition between companies. They are generated by the connections among the products and services mediated through ICT, which are built through inter-corporate co-creation. In other words, the development of products and services among companies of different industries has created new value-added through the exchange and fusion of knowledge and expertise.

In this way, the construction of a smart city is truly a "ba" creating new value through co-creation rather than competition. Thus it can be considered that if these created values relate to competitive advantage, a new focus is required—not the traditional theoretical framework of competitive advantage through inter-corporate competition but that of co-creation among companies.

Nevertheless, the focus presented in this book of building a competitive advantage by co-creation does not reject the traditional theoretical framework of building advantage through competition. The fundamental principle of competitive advantage, that it resides in the competitive environment among companies, cannot be challenged. For argument's sake, even if companies acquire a competitive advantage by co-creation among companies through the construction of smart cities, this competitive advantage will eventually be exposed to competition with other companies in the market. It follows that although the general structure of competitive advantage premised on inter-company competition does not change, existing theories relating to competitive advantage have a strong tendency toward the concept of winning through competition, and focus little on advantage through co-creation with other companies. Smart cities provide a good opportunity to demonstrate the competitive advantage of this new thinking and add a new page to the analysis of competitive advantage from a multifaceted perspective.

1.5.2 Competitive Advantage and the Creation of Social Value

Another important point from the perspective of building competitive advantage through co-creation is that of creating social value. That is to say, the value produced by through inter-corporate co-creation must have social significance. Along with the progress of globalization, competition in the market is growing increasingly fierce, and competition between companies over technological development, prices, and services is severe. In such situations, some companies with advantages in the market act with the extreme self-interest of irresponsible capitalism and in some cases do harm rather than good to society. The U.S. financial crisis of 2008 can also be said to be the result of greedy corporate management. Greedy companies may pursue competitive advantage to raise profit in the short term, but this management style is not sustainable. The reason is that management that does not confer social value cannot obtain support from the community, and eventually the customer will walk away.

The danger also exists that companies engaging in relentless competition day and night will forget to review such issues as social value creation and unwittingly fall into the trap of irresponsible capitalism. In this regard Porter and Mark (2011) advocates the idea of creating shared value (CSV). CSV means to create corporate and social value simultaneously, and aims to bring about management practices that raise a company's competitive advantage by tackling social issues. This kind of enterprise management concept has long been recognized in corporate social responsibility (CSR), but CSR focuses interest in the issue of the social responsibilities that companies should carry out, and undeniably closes its eyes to the important perspective of value-creating management. CSV tries to create a virtuous circle by reconsidering CSR from a strategic perspective. Companies create social value by tackling social issues strategically, and so build a competitive advantage together with corporate value.

The new perspective on competitive advantage presented in this book posits that co-creation among companies generates corporate and social value. As mentioned previously, the construction of smart cities is planned to tackle the social challenges of building a low-carbon society, and it has become possible for companies to create social value by tackling this issue. For that matter, the social value created by the construction of a smart city is not limited to building a low carbon society. Creation of significant social value in areas such as security, mobility, and health care are also anticipated. They can make a great contribution to the social problems that plague modern cities, which include crime, population growth, aging of society, and traffic congestion. Moreover, creating such social value enhances a company's reputation and leads to the creation of new corporate value. New corporate values go on to build a new competitive edge as the process of building smart cities with its repeated exchange and fusion of knowledge and expertise becomes incorporated into the business as tacit knowledge.

When considering the issue of a company's competitive advantage in the 21st century market, it is important to take a perspective of creating social value. We tend to associate a company's value-creating activities with the economic value of rising share prices. Of course, these values are important to corporate activity, but corporate management that leans toward economic value becomes greedy and disseminates harm in society. As mentioned above, such companies may gain a temporary competitive edge in the market, but it will not last. Looking around at the

Fig. 1.1 The focus on competitiveness

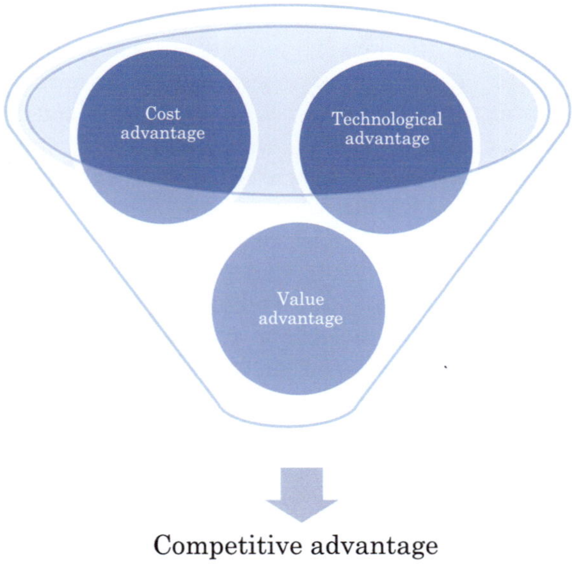

world's markets, outstanding companies that achieve a sustainable competitive advantage all have the intention to create social value. Nevertheless, when considering the problem of competitive advantage in companies in the past, the issue of social value was hardly considered. Instead, a company's competitive advantage was generally discussed in terms of issues of technological and cost advantages.

However, the society of the 21st century faces various social problems, especially global warming but also extending to the environment, population, food, and poverty, and companies should seek to tackle these problems in strategic ways. This is why the new businesses of Base of the Pyramid (BOP) and social network business have been in the spotlight in recent years. Thus, a consideration of issues of corporate competitive advantages needs to ponder not only the issues of technological and cost advantages that were the focus of previous views, but also that of value advantages creating some social worth. The focus on competitiveness presented in this book brings this area to the forefront of consciousness (Fig. 1.1).

1.6 Conclusion

Finally, I would like to consolidate the findings and issues in this book. Great anticipation surrounds the building of smart cities as policies to solve the various issues, especially global warming, that are facing modern urban areas, and building projects are ongoing in all parts of the world. This book ascertains the links between the building of smart cities and corporate competitive advantage and discusses them from the viewpoint of a new corporate advantage with two aspects: the creation of competitive edge through co-creation and building social value. As mentioned above, the previous focus on competitive advantage was generally discussed in the framework of corporate competition, and competitive advantage built through co-creation among companies tends to be overlooked. Nevertheless, smart-city construction is a platform for cross-industry cooperation among companies. The exchange and fusion of knowledge and expertise that individual companies possess creates links, and the integration of these links go to make up the smart city. This book emphasizes that when companies build a competitive edge during this process, it should be seen as coming about through co-creation among companies.

Moreover, as Porter's ideas of corporate social value show, issues of social-value creation and competitive advantage are increasingly recognized as important. Responsiveness to CSV is growing after reflection on extreme profit focus, known as "greed" in the U.S. and Europe. It goes without saying that companies are the main player in market economics, and corporate activities have a great impact on society. Contemporary society is affected by many diverse issues, especially global warming, and if companies can tackle these issues to bring about significant results they can obtain great support from society, which is certain to lead to a competitive advantage. This book focuses on building social value in this way through co-creation among companies.

As already indicated, it is important to have multiple perspectives when discussing issues of corporate advantage. Considering corporate activities from one aspect only, as if it provided the sole source of competitive advantage, would miss the overall picture. Accordingly, the discussion in this book also stands on recognition of this point. This discussion paper is one several analytical approaches to competitive advantage, and aims to present a different concept and focus to previous analytical approaches. However, this alone would not be enough. The concept has meaning when it is initiated in practice and begins to demonstrate its effectiveness. To achieve this, the ideas present in this book must build up a more specific and elaborate theoretical framework. For example, concerning the key concept of co-creation presented here, the book will describe its specific mechanisms and clarify the character of "ba," which has arrived on the stage of co-creation, while referring to previous studies. Moreover, the book must build a theoretical framework based on detailed studies of arrangements of environment and conditions that create social value through co-creation. This is the work I want to achieve in Chap. 2.

References

Aoshima, I., & Kato, T. (2003). *Strategic management.* Tokyo: Toyo Keizai Shinpou-sha.

Barney, J. (1986). Strategic factor markets: Expectations, luck and business strategy. *Management Science, 62*, 777–795.

Barney, J. (1996). *Gaining and sustaining competitive advantage*. Boston, MA: Addison-Wesley.

Campbell, T. (2012). *Beyond smart cities: How cities network, learn and innovate*. New York: Routledge.

Deakin, M. (Ed.). (2014). *Smart cities: Governing, modelling and analyzing the transition*. New York: Routledge.

Ercoskun, O. Y. (Ed.). (2012). *Green and ecological technologies for urban planning: Creating smart cities*. Hershey, PA: Information Science Reference.

Gibbs, D., Krueger, R., & MacLeod, G. (2013). Grappling with smart city politics in an era of market triumphalism. *Urban Studies, 50*(11), 2151–2157.

Hamel, G., & Prahalad, C. K. (1994). *Competing for the future*. Boston, MA: Harvard Business School Press.

Herrschel, T. (2013). Competitiveness and sustainability: Can 'smart city regionalism' square the circle? *Urban Studies, 50*(11), 2332–2348.

Itami, H., & Karube, M. (Eds.). (2004). *Strategy and logic of invisible assets*. Tokyo: Nihon Keizai Shinbun-sha.

Nikkei BP Cleantech Institute and Techno Associates. (2011). *World smart city survey 2012*. Tokyo: Nikkei BP.

Nikkei BP Cleantech Institute and Techno Associates. (2012). *World smart city report 2013*. Tokyo: Nikkei BP.

Porter, M. (1980). *Competitive strategy*. New York: Free Press.

Porter, M. (1985). *Competitive advantage: Creating and sustaining superior performance*. New York: Free Press.

Porter, M., & Mark, R. K. (2011). Creating shared value: How to reinvent capitalism—and unleash a wave of innovation and growth. *Harvard Business Review*, 1–17.

Prahalad, C. K., & Hamel, G. (1990). The core competence of the corporation. *Harvard Business Review, 68*(3), 79–91.

Rassia, S Th, & Pardalos, P. M. (Eds.). (2014). *Cities for smart environmental and energy future: Impacts on architecture and technology*. Heidelberg, New York, Dordrecht, London: Springer.

Rumelt, R. P. (1984). Towards strategic theory of the firm. In R. Lamb (Ed.), *Competitive strategic management* (pp. 556–570). Englewood Cliffs, NJ: Prentice-Hall.

Rumelt, R. P. (1991). How much does industry matter? *Strategic Management Journal, 12*(3), 167–186.

Shaw, K. (2013). Docklands dreamings: Illusions of sustainability in Melbourne docks redevelopment. *Urban Studies, 50*(11), 2158–2177.

Townsend, A. M. (2013). *Smart cities: Big data, civic hackers, and the quest for a new utopia*. New York, London: W.W. Norton and Co.

Tretter, E. (2013). Sustainability and neoliberal urban development: The environment, crime and the remaking of Austin's downtown. *Urban Studies, 50*(11), 2222–2237.

Wernerfelt, B. (1984). A resource-based view of the firm. *Strategic Management Journal, 5*, 171–180.

Wernerfelt, B. (1995). The resource-based view of the firm: Ten years after. *Strategic Management Journal, 16*, 171–174.

Chapter 2
A Theoretical Framework for Relationship-Based Strategies

In this chapter, I will present a theoretical framework for the concept of acquiring a competitive edge through co-creation. As I mentioned in the first chapter, existing theories relating to competitive advantage are structured on the assumption of corporate competition, and the competitive environment has come to be perceived as conferring a competitive advantage. In this regard, the author takes the position that corporate "co-creation" confers a competitive edge. Put another way, this is a story of creating value among multiple companies through co-creation that a company operating independently would be unable to produce, and going on to build a new competitive advantage.

But first, what kind of action is "co-creation" in the first place? In recent years, the term has come to be used in various fields, and this diverse concept is difficult to organize into a theoretical framework. In many cases the term is used similarly to refer to simple "collaboration" among organizations. In this chapter I will make a clear distinction between inter-organizational collaboration and co-creation, analyze the question of what actions "co-creation" suggests considering existing research, and formulate a clear definition.

In addition, I will examine three components—"ba," or shared context in motion, "synthesis," and "emergence"—comprising a theoretical framework for acquiring a competitive advantage through co-creation. Each component is an indispensable element for companies to create new value through co-creation. By organically linking these three elements, co-creation becomes a significant practice.

In this chapter, I will name the co-creation strategies developed from these three elements "relationship-based strategies." I will adopt these strategies as a theoretical filter when I analyze case studies of smart city construction projects from Chap. 3 onward.

2.1 What Is Co-Creation?

In recent years, the concept of co-creation has come to be used in diverse fields. Traditionally, the creative practices that gave rise to revolutionary inventions, discoveries, or ideas were thought to depend on individual qualities, and creative

© Springer Japan 2016

N. Tokoro, *The Smart City and the Co-creation of Value*,
SpringerBriefs in Business, DOI 10.1007/978-4-431-55846-0_2

acts in science, the arts, and other fields were also taken to arise from the abilities of outstanding individuals. However, in-depth observation of the actual situation made it clear that they did not arise in a vacuum.

For example, the episode where Isaac Newton, who could be called the father of modern science, conceived the notion of universal gravitation when an apple fell out of a tree is famous, but this isn't to say that he conceived the law in a flash of inspiration out of the blue. He had accumulated scientific knowledge relating to gravity and the orbits of the planets established by his scientific forebears, including Galilei Galileo and Johannes Kepler, and he established the law of universal gravitation on the basis of these intellectual assets. The phenomenon of the apple falling from the tree simply provided the spark of opportunity. Generally speaking, while the discovery of the law of universal gravitation is thought to have sprung from Newton's individual nature, reevaluating this perspective, we can also say that that the law's discovery was a product of co-creation of knowledge from Newton and his scientific forebears.

Thus, properly speaking, creative acts do not follow the pattern of a single genius creating new knowledge with no context. Either in real time or in referring back to the past, they manifest through some kind of knowledge exchange with others. This process can also apply to corporate activities. To create innovative technologies, products, service, and expertise, rather than relying on the qualities of an individual, it is important to establish mechanisms to encourage knowledge exchange among multiple agents. In recent years, interest in this perspective has increased, and a wide range of research has been undertaken on the subject of co-creation.

2.1.1 A Review of Prior Research into Co-Creation

In recent years, co-creation research in the area of service marketing has produced some striking results. In this field, the idea of Service-Dominant Logic (S-D logic), seen as the co-creation of service value with the customer, has been put forward and is greatly changing the concept of services. Proposed by Vargo and Lusch (2004), S-D logic emphasizes how the value creation model is changing from the traditional product-price based model to one that prioritizes service systems and logic.

This idea kick-started a range of research analyzing the processes and mechanisms of value creation through companies and customers, including the work from Grönroos and Voima (2013), Echeverri and Skälen (2011), Payne et al. (2008), Vargo et al. (2008), Sssrijärvi (2012), Mele et al. (2010), Karababa and Kjeldgaard (2014), and Gummerus (2013). All of this research focuses on the mechanisms and processes of co-creation among companies and customers, emphasizing the viewpoint of co-creation among agents creating value. Nevertheless, this research focuses on analyzing the processes and mechanisms of value co-creation, and does not clearly define the practice of co-creation itself.

Prahalad and Ramaswamy (2003, 2004a, b) provide research that can be viewed from the perspective of corporate competitiveness through value co-creation among company and customers. They emphasized that corporate competitiveness in 21st-century markets is brought about through value co-creation and introduced the idea of co-creation in the strategic domain. According to these authors, companies were the agents of value creation in markets up to the 20th century, and the customers' role was only to compare value created by the company and to pay.

In contrast, in the 21st century, marketplace value is created through co-creation among corporations and customers, and the markets function as "ba," shared context in motion, for the buying and selling of products and services as well as for value co-creation between companies and customers. Naturally, corporate and customer theories of value creation differ. To discover the point of contact, co-creation experiences between company and customer have to accumulate. Within this process of accumulation, dialogue is important.

Naturally, the dialogue I am referring to is not simple conversation. Rather, it is productive, creative dialogue linked to the process of creating of new value. Prahalad and Ramaswamy noted that it was necessary to create rules for participation in order to achieve productive, creative dialogue in an orderly way. Their research introduced the new concept of co-creation in strategic domains, and brought fresh ideas to the existing thinking on corporate competitiveness. However, their concerns were limited to questions of creating value with customers and did not deal with, for example, the co-creation of values among companies in different industries, which is the analytical aim of this chapter. Moreover, without a clear definition of the practice of co-creation, they are also vague on how co-creation differs from conventional inter-organizational collaboration.

Research on co-creation is not limited to the domain of social science. In recent years, interest in clarifying the mechanisms behind the practice of co-creation in the natural science domain has also grown. For example, in 2004 the University of Tokyo's RACE (Research into Artifacts, Center for Engineering) set up a co-creation department, and is taking an engineering approach to analyzing the mechanisms of co-creation. According to Ueda (2004) and others, the aim of co-creation engineering is to change the state of the co-creative relationship between human beings and artificial objects. This approach methodology is not the top-down analytical approach of control theory and optimization theory that is orthodox in the engineering field but the bottom-up method with the goal of constructing a theory. Research is ongoing in the engineering field, and the definition of co-creation practices is being clarified.

In more detail, the practice of co-creation in co-creation engineering is being defined as "the practice of creation as a general system obtaining results from interaction among active agents for solutions that cannot be obtained through active agents acting in isolation alone." The subjects of the action are not limited to human beings but also include the organizational bodies of intelligent artificial bodies, corporations, and others. Co-creation engineering is not about clarifying the mechanism of mutual action through this kind of actor agent, taking an analytical approach (analyzing in detail the various elements behind the construction of a system, checking their qualities, and thereby grasping the system as a whole) but should be understood as the partial interaction of various elements that by and by come to control the system as a whole and lead to new creation. This kind of analytical approach could be termed “synthesis.”

2.1.2 Defining Co-Creation

A review of prior research in recent years shows great activity in research relating to co-creation in various fields, but it would be hard to say that a clear definition of co-creation has been established. Provisionally, if the action of people cooperating together in some kind of work is understood as a co-creative practice, then co-creativity is happening in places where it reaches society. Moreover, concerning relationships among companies, if inter-corporate collaboration in the form of technology links and sales cooperation is recognized as co-collaboration, then co-collaboration is not a special, original practice, but something quite mundane. It could also be seen as inter-corporate co-creation including open innovation, as espoused by Chesbrough (2003a, b, 2006).

Looked at like this, co-creation can be considered not as a special, innovative concept but as a common practice in society. Without a clear definition of co-creation, it is difficult to distinguish it from these existing practices. Nevertheless, the following will show that the co-creation hypothesized in this chapter is clearly distinct from this. “Co-creation is a practice of creating value through cooperation among multiple active agents with certain shared purposes that agents acting in isolation cannot achieve.”

Put another way, the “co-creation” assumed in this chapter does not indicate simple cooperative relationships among individual actors. For example, if a 30 kg. weight were divided among three people, each person would assume a 10 kg. burden. Could the result of that action be termed “co-creation”? No, for the reason that these three individuals are not creating value but simply undertaking their task through a division of roles. The same logic applies to collaboration among companies. As I mentioned previously, the various forms of collaboration undertaken among companies, whether technology links, sales cooperation, or open innovation, cannot be termed “co-creation” if they are based on simple divisions of roles among companies and do not create new value.

Concerning co-creative practices, let’s consider soccer as a metaphor of a mass game. Soccer is played with 11 players but with the exception of the goalkeeper, the division of labor among the players is blurred. There is a general separation of offense and defense, but within the game’s flow each player is required to respond to changing conditions. If they see an opportunity, the players whose role was originally defense can join the attack, pressing to score against the opposing team by sallying forth in a variety of attacking patterns. Within this process, various creative plays are made that thrill the spectators. Soccer could be termed a very co-creative sport. The more the players work as a team, the higher the level of co-creativity among the individual players. In this way it is different from mass gymnastics, which is another group sport.

In the case of mass gymnastics, the division of individual roles is strict, and the performance plays out exactly as planned. Individuals are not permitted to respond to conditions with creativity and originality. As a result, they display loyalty to their predetermined role with comprehensively controlled performances. These games show collaboration among actors but are not co-creative. Understanding co-creation in this way, I will consider, in order, the following three components involved in realizing co-creation.

2.2 The Presence of "Ba"

"Ba," shared context in motion, is an important concept when considering the process of value being created amid interaction among different actors. Original research into "ba" can be traced back to the concepts of electrical and magnetic fields in 19th-century classical physics. "Electrical fields" are the condition of space surrounding electrically charged objects, and magnetic fields are created by electrical current. That magnetism indicates the condition of that magnetism acting on other electrical currents.

In classical physics, this kind of "ba" basically comprises physics-related entities, and was thought to be unrelated to objects. Later, in the 20th century, Einstein developed the theory of relativity, noting that matter and fields cannot be separated, and emphasized that where matter exists, a gravitational field also exists. According to Heisenberg (1958), in modern physics matter does not exist in isolation, but is indivisible from its surroundings. He grasped that the characteristics of matter are determined by its relationship with its surroundings.

Meanwhile, in the field of sociology, the research into gestalt social psychology of Lewin (1951)[1] is applicable. He defined the fact of coexistence considering general mutual dependence as "ba," and thought it necessary to consider people and the lifestyle space that they inhabit as a single "ba" in psychological terms. Moreover, Nishida (1965) and Shimizu (2000, 2003) contributed with research in the field of philosophy. These people developed a theory of unique "ba" based on the concepts of the virtue of integrating the agent and object, or inseparability of the agent. Such concepts from Eastern philosophy also influenced the knowledge creation theories advocated by Nonaka (1995, 1998), and the concept of "ba" in the management domain was fully introduced. Hiroyuki Itami (1999) also contributed to research into "ba" in a management context.

In this way, research into "ba" has been carried out in a variety of fields, and recognition of the importance of such research in management is growing. In other

[1] A collection of Lewin's posthumous manuscripts edited by Cartwright and published in 1951.

words, as mentioned previously, amid the exchange of different actor agents and interaction incorporating a range of knowledge, expertise, and technology, the process by which new value is created can be considered to take place on the stage of "ba." In these cases, "ba" does not only indicate physical areas such as offices and meeting rooms. For example, cases of information exchange through the internet among people separated geographically by some distance and not known to each other are recognized as "ba" of information exchange. By writing and other means, the actors can grasp ideas in the same way through psychological ties as "shared 'ba' of thinking." In other words, "ba" is an extensive relationship concept including physical space, virtual space, and shared mental space among people.

So what conditions are required of "ba" in order to achieve co-creation? One is the existence of some kind of shared purpose among participating active agents. Simply bringing together independent active agents will not lead to interaction. The "ba" of a station is a place where numerous people interact, but these people have no shared purpose. People who board a train for the purposes of commuting to work or school use the station as a place to rendezvous, or to buy from a station shop. They have various purposes, and just happen to be present at that place. In such cases, the exchange of mental energy and knowledge among people participating in a "ba" does not take place, nor is any value created. In other words, this place (which is also pronounced "ba" in Japanese) has no meaning other than that a large number of people are present at the same place.

In contrast, a shared purpose exists among people taking part in a gathering at a plaza in front of the station. The meeting has an agenda, and the people taking part have some kind of purpose and interest regarding the agenda. They are participating in this "ba." In this case, the participants exchange mental energy through being sympathetic, antagonistic, or angry, and the exchange of knowledge is encouraged through interchange of opinions among participants. As a result, some kind of message is developed through "ba" with the potential for creating value.

Another condition is the activation of "ba" and the presence of the kind of "management ba" that stimulates co-creation. Among active agents participating in "ba," even if a shared purpose exists, that does not mean that new value is constantly created. What's more, we can safely assume that productive, creative dialogue does not somehow arise simply by adjusting the interests among active agents. We can consider that the clash of competing interests among active agents can deepen, and lead to an irreparable situation. When "ba" falls into this kind of condition, we begin to doubt not only the creation of value but also the continued existence of "ba."

In fact, on these points, past research analyzing collaboration among organizations bears this out.[2] For example, the research of Vangen and Huxham (2003),

---

[2]This chapter makes a clear distinction between "co-creation" and "collaboration," but refers to this previous research from the viewpoint of proceeding with tasks through cooperation among different organizations.

shows how numerous organizations co-create value by collaborating with other organizations, hoping to achieve excellence as a result, but in fact this is extremely difficult to manage, and success or failure depends on the ability to build inter-organizational trust. Moreover, research from Kanter (1994), Dacin et al. (1997), Gray (1985), Wistow and Hardy (1991), and others, while accepting the importance of creating value through inter-organizational collaboration, points to the difficulty of collaboration process management.

So what kind of "ba management" is necessary to activate ba vigorously, bring about creative interaction among participating active agents, and create value? Hiroyuki Itami mentions the following five conditions.

(1) Stirring up (or fluctuation)
(2) Clearing up debris
(3) Creating a path
(4) Creating a flow
(5) Finishing off

(1) Stirring up is to be stimulated by "ba." This involves breaking down the existing order and values, and taking the opportunity to create new flows leading to fluctuations in the organization.
(2) Clearing up debris involves discovering flashes of insight among people who have started to perform new actions by stirring up.
(3) Creating a path involves perceiving directions indicated by clearing away diverse debris and consolidating with appropriate expression. However much debris is cleared up, an organization cannot find a direction while some is still scattered about. It is necessary to understand the values lying in the deep layers of debris and merge them.
(4) Creating flow involves giving a supportive push to get people leaning toward creating a path to move independently.
(5) Finishing off involves stopping regularly so that everyone can confirm the directional flow. Even if people start moving in the direction of flow, the flow can become wayward unless that work is undertaken regularly, and can move in unfavorable directions. Regularly "finishing off" is necessary to confirm the directional flow.

The duties of such management "ba" are undertaken by the "ba" leader. Outstanding leaders that can accomplish "ba management" in order to manage various actor agents with different backgrounds and interests and realize value co-creation are indispensable. However, the abilities required are clearly different from those needed by the leaders of bureaucratic organizations. Leadership dependent on power and authority will cause "ba" to wither. The abilities required of the leaders who take on the duties of "ba management" are different from those needed in conventional organizations. It is not easy to cultivate "ba" leaders. The only way is to go through a repeated process of trial and error and to accumulate experience. At these times, it is important to cultivate people who support the leaders (Fig. 2.1).

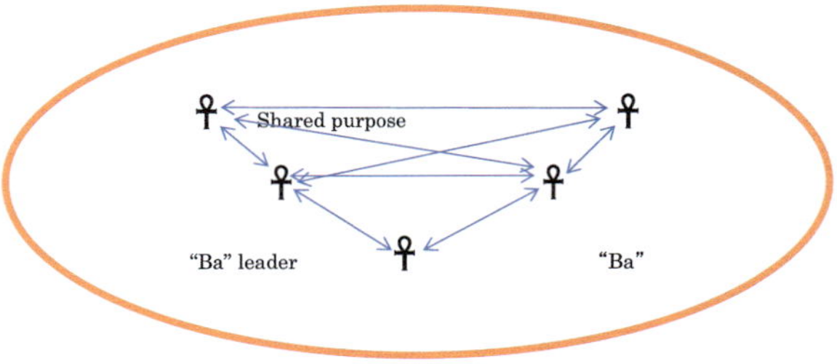

Fig. 2.1 Conceptual diagram of "Ba"

2.3 Synthesis

The second component for achieving co-creation is "synthesis." Synthesis is a scientific inquiry methodology grasped conceptually with the meaning of general integration or unification, and generally understood in the field of natural sciences to be comparable to "analysis."

In the world of the natural sciences, the analytical approach has been tried for everything from the origins of the universe to the mechanisms of the human body. Analysis is reductionism, and it is an approach method for comprehending multiple systems. It attempts to do this by grasping the nature of the individual elements and the relationships between them by dismantling the system element by element, and so understanding the system as a whole. Humanity has attempted to come to grips with the multiple mechanisms of the natural world by using the analytical method, and little by little, over a long period, it has come to grasp those systems.[3] Put another way, the scientific approach is presented through analysis, and in recent years this approach has been incorporated even into the area of social science. For example, in order to pursue the facts in the background of some social phenomena, typical methods include trying to analyze the data statistically and grasp the quality of and relationships among data.

In contrast, "synthesis" indicates an opposite vector to "analysis" in the analytical approach. Where analysis dismantles the individual elements of a system, the synthesis approach combines each element to build a system. Put another way, analysis acts from the whole to subgroup while synthesis proceeds from subgroup to whole (Fig. 2.2).

But while analysis and synthesis are completely opposite approaches, they cannot be said to have no mutual relationship. For example, when analyzing natural phenomena, the analytical approach that is undertaken incorporates synthesis. After analyzing the individual elements of phenomena, the process of recombining them to understand the phenomena is always carried out. In contrast, even with some kind of creative process where the dismantled elements are recombined, the process

[3]Meanwhile, it is certain that the natural sciences have come to select only those targets that are capable of being analyzed. In other words, reductionism excludes those targets that resist analysis.

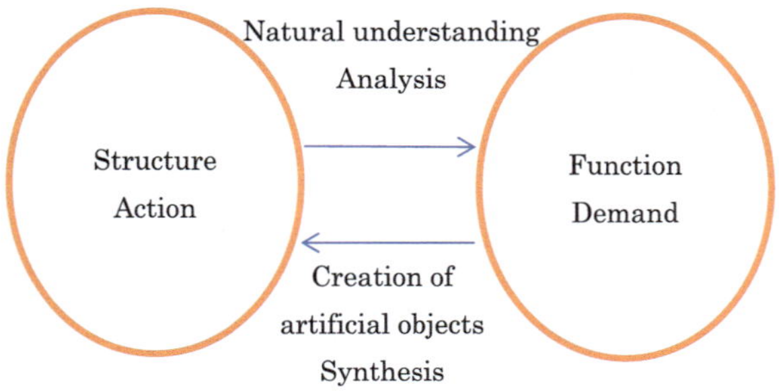

Fig. 2.2 Analysis and Synthesis. *Notes* Natural understanding: analysis through synthesis (wanting to understand multiple parts). Creation of artificial objects: synthesis incorporating analysis (wanting to create multiple objects). *Source* Ueda (2004) compilation. *What is Co-Creation?* Baifukan, p. 44

of combination requires that the individual elements be analyzed. In this way, analysis and synthesis are actually related in multiple ways, and whichever approach is taken, one cannot be said to bear no relation to the other.

In order to achieve co-creation among active agents, the active agents must implement synthesis. The reason is that the act of co-creation cannot arise from the analytical approach. In other words, the creative act of co-creation is not understanding by breaking down individual elements and rebuilding them, but rather the act of combining various elements and, while integrating them, building some kind of artificial creation. Let's consider the example of building a smart city. In order to understand the mechanisms by which environment, mobility, health care and other factors create value for the smart cities that already exist, it is necessary to take an analytical approach. By breaking down the various systems that comprise the smart cities into individual elements and examining the functions of each element and the inter-element mechanisms, we can try to understand the mechanisms that create value.

Set against this, to construct a smart city under a shared purpose, a synthesized approach is necessary in order to analyze the series of processes by which various companies create new value while mixing together and fusing technology, know-how, and knowledge through participating in "ba," and so build the creation of a smart city population. More succinctly, a synthesized approach is required to understand the mechanisms by which different actor agents exchange and merge knowledge to create value. For example, research by Nonaka et al. (2006) analyzes the processes whereby individual knowledge merges with organizational knowledge through the synthetic approach. With this kind of thinking, we can consider that co-creation is truly synthetic, and that the analytical approach through synthesis can help us to understand the mechanisms of every creative act.

However, what we have to keep in mind here is the point that synthesis is an interpretative approach that involves the analyzer's subjectivity. Let's take another look at the example of the smart city given above. The co-creation mechanisms involved in the creation of smart cities and the processes of knowledge exchange and fusion among active agents would yield the same results whoever was performing the analysis. Under the same conditions, you would expect to find the same results for other projects. If this is so, then it would be possible to hypothesize a general universal law on a scientific basis from these results. However, in reality this does not happen. The mechanisms of co-creation are obtained from the analysts' range of results, and different projects yield different kinds of knowledge. The reason is that the analysts focus on subjective interpretative theories.

Generally, with the scientific analytical approach, it is important that the analysts exclude subjectivity and pursue objective facts alone. Regarding results obtained from experience and observation, facts can be acknowledged as objective as long as the same facts are obtained by whoever investigates them. Any analytical approach that yields different results cannot be termed scientific. Accordingly, with the scientific approach, objectivity-focused methodologies of deductive and inductive reasoning are required.

Seen from such a standpoint, the analysts' subjective-based interpretative approach cannot be said to be scientific, since analysis through synthesis cannot be termed scientific. On this point, the writer's opinion is clear. In other words, when viewed from a scientific context from the viewpoint of the natural sciences, the interpretative approach is unscientific, but if the definition of science were established differently to that of dependence of natural science, the picture changes. Essentially, the objects of natural and social phenomena targeted by the natural science and social science approaches respectively differ, and the same standards are not applicable to both.

With the natural sciences, whose object is to clarify the mechanisms of natural phenomena, it is possible to establish universal laws by excluding subjectivity and pursuing the facts objectively, but in the social science domain this is very difficult to achieve.[4] In the case of natural phenomena, although many areas have yet to be explained, fixed laws for the natural world do exist, and those laws are seen to be held universally. In contrast, it is difficult to discover universal laws among social phenomena, which comprise a collection of actions brought about by people with opinions that change in response to the situation. For example, human beings have fought each other since the dawn of history, and if a universal law could be provisionally formulated to explain the mechanisms by which wars occurred, it would become possible to prevent war. Yet such a law does not exist.

Given this reality, how should the scientific element of social science be defined? We certainly should not append the term "science" to a situation where multiple advocates arbitrarily develop their theories according to their own subjective ideas. On this point, I would stress the following: in the social sciences (especially in the

---

[4]Even in some fields of the social science domain, such as economics, there is a tendency to take an approach based on natural science models and pursue the formulation of universal laws through mathematical models.

writer's specialism of management studies), while it is difficult to formulate universal laws from the results of observing and analyzing social phenomena, it is possible to fathom the logic behind those phenomena.

However, such logic has specific characteristics for applicable social phenomena, and will not work on other phenomena. Nevertheless, studying this logic will help to make analysis of other social phenomena more accurate. It follows that the analytical approach through interpretative theories requires that the analyst understand the logic behind the social phenomena targeted for analysis and create an abstract working process. With such a process, social science can be understood as a science.

2.4 Emergence

The third component in realizing co-creation is "emergence," a type of action that manifests bottom-up. The term might apply, for example, to a series of processes by which a minor activity that at first only occurs at the fringes of an organization grows as time goes by until it develops to drive the entire organization. Accordingly, this action is the opposite of top-down movements.

The first person to espouse the concept of "emergence" in management studies was Mintzberg. The field of management studies, especially management strategy theory, had come to see the conception and planning of strategy as work that should be undertaken by the upper echelons of top management, who should create systematic plans indicating an organization's future direction considering the various internal and external conditions surrounding the organization in a top-down manner (Ansoff 1965; Andrews 1971; Steiner 1969) . Against this, Mintzberg stressed that rather than conceiving and planning strategy top-down in advance, it should take shape at the practical level of the workplace through a process of trial and error (Mintzberg 1973, 1978, 1990).

Mintzberg analyzed, in chronological order, Volkswagen strategies from 1920 to 1970 and American strategies for Vietnam from 1950 to 1973. He emphasized that the strategies were planned to begin with (intended strategy), and as time went by this intended strategy split into unrealized and deliberate strategies. Emergent strategy appeared part-way along the timeline and combined with deliberate strategy to become realized strategy. He believed that the limitations of human cognitive capacity make it impossible for a specific set of people to accurately grasp and plan the entire situation in advance, and that in reality, strategy was implemented differently to the originally planned intended strategy. Because of this, it is important for organizations to have the freedom for strategic ideas to bubble up from the middle and lower levels of the workforce and so activate hidden growth potential (Bower and Gilbert 2007; Quinn 1978, 1980; Burgelman 1983, 1994).

The way of thinking about these strategic theories leads to favorable suggestions when considering co-creation practices. For example, the Japanese automaker Honda has adopted a unique initiative known as "waigaya," a term that refers to informal discussions where staff can express themselves freely without regard for position, seniority, or gender. Honda's "waigaya" usually happen over a period of several days in a place chosen to be outside the daily routine, such as a hotel or recreational accommodation. Staff discuss issues freely while eating and sleeping at the venue.

The topics of discussion are not specific issues such as how to raise sales and profit rates, but very abstract, substantive issues such as, "What is a car that people love?" Accordingly, "Waigaya" does not always lead to results, nor can the company expect results every day. Sometimes it becomes no more than an informal social gathering. Yet Honda perseveres with Waigaya despite the company footing all the bills. It does this not to convey innovative ideas and concepts top-down in a managed format, but rather because of the high possibility that something will emerge from the interactive exchanges among staff through "ba" in an atmosphere of greater freedom.

Under the "Waigaya" system, in many cases the first day of the stay is formal, and numerous participants take part in superficial discussions, but as the second, third, and more days go by, the atmosphere becomes less reserved, and in many case more substantive discussions develop. With "waigaya," participants are guaranteed the right to express themselves freely regardless of position, seniority, or gender, and care is taken that what is expressed during these discussions is not held against the employee. Although results are not guaranteed, this atmosphere of freedom draws out the real feelings of the participants, and amid the clash between this expression and the true feelings that cannot be shared at the office on a daily basis, new ideas and concepts take shape. This series of processes is truly co-creation through emergence.

Moreover, another Japanese company, the electronics manufacturer Canon, implements a "morning meeting" system where directors meet at 8 a.m. for a discussion with no set topic. This "morning meeting" differs from the usual directors' meeting due to its informality. Accordingly, the discussions that take place at these meetings do not directly influence corporate activity. The employees freely discuss a wide range of issues, from topics of the day to the weather, and exchange their opinions. The purpose of the "morning meetings" is to cultivate mutual trust among employees, but among these unstructured discussions new concepts are born that may influence Canon's strategy. Honda and Canon both create a succession of outstanding products, and both companies are deeply interested in having a system that can give rise to "emergence."

Here I would like to consolidate the features of "emergence." The first feature is the flow of activity from bottom to top of an organization known as "bottom-up." This is the process by which activity at the workplace of corporate organizations expands through interaction among employees and percolates up to the senior echelons in a more evolved form. It is quite different from the decision-making process undertaken by conventional organizations, which are typically more

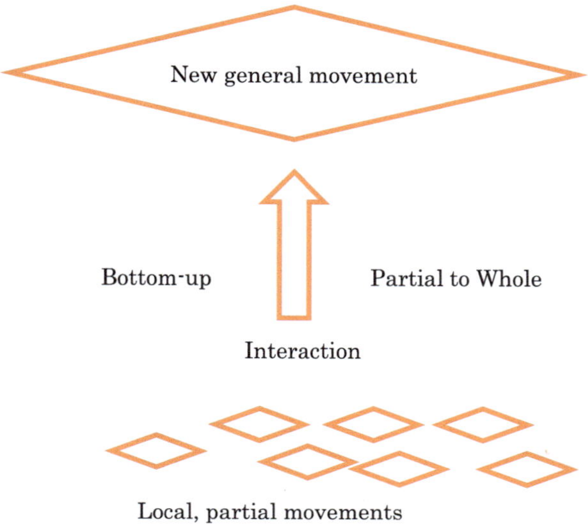

Fig. 2.3 Conceptual diagram of emergence

bureaucratic. As touched on by the strategy theories discussed above, at conventional organizations, the senior echelons that hold the authority often create plans and implement them in a top-down manner, but this modus operandi does not give rise to "co-creation" among employees. With the top-down method, employees are only permitted to act within the boundaries of predetermined plans, and each person's role is limited. Accordingly, to encourage co-creation among employees, it is necessary to assure a bottom-up rather than top-down flow of activity.

The second feature concerns the flow from partial to whole. This has similarities to the first feature, in which actions flow from bottom to top. In the initial stage, an organization's local activities can be seen at a fringe level, and amid interaction among active agents, a number of localized movements consolidate to form a greater flow, and eventually develop to embrace the whole organization, having an effect on the organization's character and activities. In other words, rather than having a rough idea of the whole from the beginning and breaking it down, the organization creates a structure that builds activities from partial to whole over time. The series of processes that builds from partial movement to the whole is implemented through interaction among actor agents.

Finally, the third feature is an organization's degree of freedom. The greater the freedom, the higher the chance that emergence will appear. Put another way, the more open and "well-ventilated" the organization, the easier it is for emergence to appear. As seen in the case of Honda's Waigaya mentioned above, emergence can easily appear in an organization that nurtures a climate where anyone can express themselves freely regardless of organizational level, age, or gender. In contrast, emergence can hardly be expected in organizations with rigid strata where it is difficult for those at the edges and in the middle strata of the organization to exchange opinions (Fig. 2.3).

2.5 Relationship-Based Strategies as Co-Creation Strategies

The three essential components to achieve a competitive edge through co-creation among actor agents are "ba," "synthesis," and "emergence." These three components do not exist in parallel, but rather have a multilayered structure. When a co-creation strategy breaks down these components into their elements through an analytical approach, the presence of "ba" constitutes the foundation of this strategy. If the "ba" that the various actor agents participate in does not exist, co-creation strategies will not be established in the first place. However, as mentioned above, the presence of "ba" alone is meaningless. Some kind of shared purpose among participating actor agents is required, as is coordinating the interests of the participating actor agents and a "management ba" that encourages the exchange of knowledge. Innumerable "ba" with a wide range of actor agents participating under a shared purpose exist in society at large, but only in some cases will the interaction among them lead to new ideas and concepts or the creation of revolutionary products and services. The qualitative difference of "management ba" lies in the background.

In this way, the presence and quality of "ba" establish the roots of co-creation strategies, and can be said to form the essential connection between "ba" and "co-creation." The other two components of synthesis and emergence have a great effect on the presence of "ba." Even if "ba" exists temporarily, in cases where there are problems with its character, the elements of synthesis and emergence must be incomplete. For example, if the shared purpose set as "ba" is unclear, the actions of each actor agent will be consistently deficient, and co-creation become difficult to achieve. Moreover, when appropriate "management ba" does not exist, we run into the danger that repeated clashes of interests among actor agents will render "ba" meaningless, and if certain actor agents operate a "ba" with excessive authority, it will be difficult for bottom-up emergence to appear.

While it follows that appropriate "ba" are assumed to exist in synthesis and emergence, what is then manifested through "ba" in these situations is emergence. As mentioned above, emergence is a series of processes by which local movements at the fringes of an organization become a major movement that drives the entire organization through interaction among actor agents from the bottom up. Models that break down movements from top to bottom will not always fail to lead to co-creation, but a situation where the division of roles is determined by a plan drawn up in advance, though it may create co-creation temporarily, is very likely to develop in a limited way.

Even though various kinds of emergence can arise through "ba," these are not directly connected to co-creation. In other words, as mentioned above, co-creation indicates creative action through cooperation among multiple actor agents that these actor agents could not achieve in isolation, but co-creation will not be seen until new value is created as an organization. Thus even if emergence occurs everywhere that is reached by "ba" and previously unseen ideas and concepts arise, they will have no meaning unless they are connected to value creation. Then the role of connecting what arises from emergence to new value creation is carried out by synthesis. The act of synthesis scrutinizes the various elements that arise from emergence and combines them to create new value.

An open "ba" with a high degree of freedom can be considered enough for emergence to occur through interaction among participating actor agents. However, this emergence is not necessarily well connected to value creation. New ideas occurring at the workplace level that lead to greater work efficiency in part of the company may not lead to new value for the company as a whole. Accordingly, synthesis is required to extract the essence connected to value creation hidden within various forms of emergence and connect them.

High-level value creation occurs not by connecting everything but by selecting those elements related to value creation and integrating them effectively. For synthesis to play this role, a blueprint is needed as to what kind of value a company wants to create. Unless a company can indicate in advance what it wants to do and what value it wants to create, it will be unable to extract the essence from emergence. This thinking is known as abduction. As mentioned previously, among methods of scientific pursuit, analysis uses the methodologies of deduction and induction while synthesis uses abduction (a logical argument whose major premise is certain but whose minor premise is probable).

In this chapter, I have called the co-creation strategies arising from "ba," "synthesis," and "emergence" relationship-based strategies. These relationships can take place among actor agents or within society. The relationships among actor agents have already been noted, but the concept implies relationships based on co-creation that creates value and builds a competitive advantage rather than relevant relationships limited by competition among actor agents. For example, as seen in the thinking of game theory, the relationships among companies in a market are generally considered to be either competitive or cooperative (Brandenburger and Nalebuff 1996; Ghemawat 1997; McAfee 2002). Moreover, standard theories on competitive advantage, such as the positioning view (Porter 1980, 1985) and the resource-based view (Wernerfelt 1984; Rumelt 1984; Barney 1986), assume that corporate competition creates competitive advantages. While these ideas are dominant, the focus on relationships predicated on co-creation is the unique focus of this book, and can be said to be different from previous perspectives on relationships among actor agents.

Another relationship that exists is that with society. This point was mentioned in Chap. 1, but the value creation through co-creation assumed in this chapter must have significant value for society. Value created on the basis of self-interest or greed, even though this value may be significant for the person concerned, can only be harmful for society. Typical of this would be the malicious development of financial products that drive people to bankruptcy. It follows that values predicated

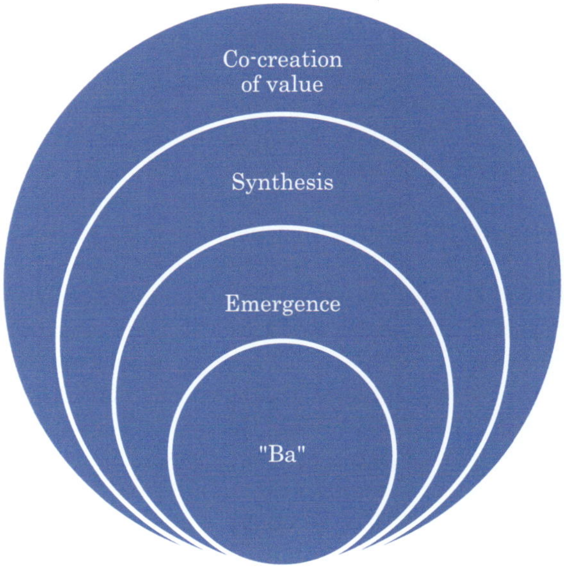

Fig. 2.4 Positioning of "ba," "emergence," and "synthesis" for relationship-based strategies

on relationship-based strategies must have social value that emphasizes relationships with society. Generating social value through co-creation is linked to acquiring societal trust, enhancing the reputation of the enterprise, and acquiring a competitive advantage (Fig. 2.4).

References

Andrews, K. R. (1971). *The Concept of Corporate Strategy*. Homewood, III: Dow Jones-Irwin.

Ansoff, H. I. (1965). *Corporate Strategy: An Analytical Approach to Business Policy for Growth and Expansion*. New York: McGraw-Hill.

Barney, J. (1986). Strategic factor markets: Expectations, luck, and business strategy. *Management Science, 32*(10), 1231–1241.

Bower, J. L., & Gilbert, C. G. (Eds.). (2007). *From Resource Allocation to Strategy*. Oxford: Oxford University Press.

Brandenburger, A. M., & Nalebuff, B. J. (1996). *Co-opetition*. New York, NY: Currency.

Burgelman, R. A. (1983). A process model of internal corporate venturing in the diversified major firm. *Administrative Science Quarterly, 28*(2), 223–244.

Burgelman, R. A. (1994). Fading memories: A process theory of strategic business exit in dynamic environments. *Administrative Science Quarterly, 39*(1), 24–56.

Chesbrough, H. (2003a). The era of open innovation. *Sloan Management Review, 44*(3), 35–41.

Chesbrough, H. (2003b). Open innovation: How companies actually do it. *Harvard Business Review, 81*(7), 12–14.

Chesbrough, H. (2006). *Open Business Models: How to thrive in the New Innovation Landscape*. Boston, MA: Harvard Business School Press.

Dacin, T., Hitt, M., & Levitas, E. (1997). Selecting partners for successful international alliances: Examinations of US and Korean firms. *Journal of World Business, 32*(1), 3–16.

Echeverri, P., & Skälen, P. (2011). Co-creation and co-destruction: A practice-theory based study of interactive value formation. *Marketing Theory, 11*(3), 351–373.

Ghemawat, P. (1997). *Games Business Play: Cases and Models*. Cambridge, MA: The MIT Press.

Gray, B. (1985). Conditions facilitating interorganizational collaboration. *Human Relations, 38*, 911–936.

Grönroos, C., & Voima, P. (2013). Critical service logic: Making sense of value creation and co-creation. *Journal of the Academy of Marketing Science, 41*(2), 133–150.

Gummerus, J. (2013). Value creation process and value outcomes in marketing theory: Strangers or siblings? *Marketing Theory, 13*(1), 19–46.

Heisenberg, W. (1958). Physics and Philosophy. New York: Harper Torchbooks.

Itami, H. (1999). *Management of "Ba": A New Management Paradigm*. Tokyo: NTT Publishing.

Kanter, R. (1994). Collaborative advantage: Successful partnerships manage the relationships, not just the deal. *Harvard Business Review, 72*(4), 96–108.

Karababa, E., & Kjeldgaard, D. (2014). Value in marketing: Toward sociocultural perspectives. *Marketing Theory, 14*(1), 119–121.

Lewin, K. (1951). *Field Theory in Social Science: Selected Theoretical Papers, Edited by Dorwin Cartwright*. New York: Harper.

McAfee, R. P. (2002). *Competitive Solutions: The Strategist's Toolkit*. Princeton, NJ: Princeton University Press.

Mele, C., Spena, T. R., & Colurcio, M. (2010). Co-creating Value Innovation through Resource Integration. *International Journal of Quality and Service Science, 2*(1), 66–78.

Mintzberg, H. (1973). Strategy-making in three modes. *California Management Review, 16*(2), 44–53.

Mintzberg, H. (1978). Patterns in strategy formulation. *Management Science, 24*(9), 934–948.

Mintzberg, H. (1990). The design school: Reconsidering the basic premises of strategic management. *Strategic Management Journal, 11*, 171–195.

Nishida, K. (1965). *A Full Collection of Kitaro Nishida's Work*. Tokyo: Iwanami Shoten.

Nonaka, I., & Takeuchi, H. (1995). *The Knowledge Creating Company*. New York: Oxford University Press.

Nonaka, I., & Konno, N. (1998). The concept of 'Ba': Building a foundation for knowledge creation. *California Management Review, 40*, 40–54.

Nonaka, I., Krogh, G., & Voelpel, S. (2006). Organizational knowledge creation theory: Evolutionary paths and future advances. *Organization Studies, 27*(8), 1179–1208.

Payne, A. F., Storbacka, K., & Frow, P. (2008). Managing the co-creation of value. *Journal of the Academy of Marketing Science, 36*(1), 83–96.

Porter, M. E. (1980). *Competitive Strategy*. New York: Free Press.

Porter, M. E. (1985). *Competitive Advantage: Creating and Sustaining Superior Performance*. New York: Free Press.

Prahalad, C. K., & Ramaswamy, V. (2003). The new frontier of experience innovation. *Sloan Management Review, 44*(4), 12–18.

Prahalad, C. K., & Ramaswamy, V. (2004a). Co-creation experience: The next practice in value creation. *Journal of Interactive Marketing., 18*(3), 5–14.

Prahalad, C. K., & Ramaswamy, V. (2004b). *The Future of Competition, Co-creating Unique Value with Customers*. Boston, MA: Harvard Business School Press.

Quinn, J. B. (1978). *Strategic Change: Logical Incrementalism* (pp. 7–21). Fall: Sloan Management Review.

Quinn, J. B. (1980). *Strategic for Change: Logical Incrementalism*. Englewood Cliffs, NJ: Irwin.

Rumelt, R. P. (1984). Towards a strategic theory of the firm. In L. R. Boyden (Ed.), *Competitive Strategic Management* (pp. 556–570). Englewood Cliffs, NJ: Prentice-Hall.

Saarijärvi, H. (2012). The mechanisms of value co-creation. *Journal of Strategic Marketing, 20*(5), 381–391.

Shimizu, H. (2000). *"Ba" and Co-Creation*. Tokyo: NTT Publishing.

Shimizu, H. (2003). *The Concept of "Ba"*. Tokyo: University of Tokyo Press.

Steiner, G. A. (1969). *Top Management Planning*. New York: Macmillan.

Ueda, K. (Ed.). (2004). *What is Co-Creation?*. Tokyo: Baifukan.

Vangen, S., & Huxham, C. (2003). Nurturing collaborative relations: Building trust in international collaboration. *The Journal of Applied Behavioral Science., 39*(1), 5–31.

Vargo, S. L., & Lusch, R. F. (2004). Evolving to a new dominant logic for marketing. *Journal of Marketing, 68*, 1–17.

Vargo, S. L., Maglio, P. P., & Akaka, M. A. (2008). On value and value co-creation: A service system and service logic perspective. *European Management Journal, 26*(3), 145–152.

Wernerfelt, B. (1984). A resource-based view of the firm. *Strategic Management Journal, 5*, 171–180.

Wistow, G., & Hardy, B. (1991). Joint management in community care. *Journal of Management in Medicine, 5*, 40–48.

Chapter 3
Co-Creation of Value Generated by a Self-motivated "Ba"—A Case Study of the Yokohama Smart Community

3.1 Introduction

In this chapter, I will analyze initiatives of the Yokohama Smart Community, which is currently conducting demonstration experiments of a smart city project aimed at achieving a low-carbon society. Promoting the development of an energy system "learned from a view of life found in nature," the Yokohama Smart Community formed a consortium comprised of participating companies, universities and local government to put into practice a range of innovative initiatives. A characteristic feature of this consortium is the absence of a specific leader. This is because the Yokohama Smart Community is not a consortium formed by a particular leader to promote a specific initiative to be implemented under that individual's or organization's leadership. Moreover, while the local government of the city of Yokohama is involved, it is purely in an advisory capacity, so the Yokohama Smart Community does not have the character of a government-led project. Therefore, the participating members voluntarily take part in the consortium as they wish without being subject to any explicit or implicit restrictions or constraints of another organization, and they are free to withdraw at any time.

The objective of this chapter is to define what co-creation of value among participating members in the context of such a self-motivated "ba," or shared context in motion, means and to shed light on this process.

3.2 Analytical Framework

Before beginning the discussion, I would first like to confirm the theoretical framework that will be used in the case study analysis here. As I mentioned earlier in Chap. 2, "relationship-based strategies" will be the theoretical framework I will refer to when conducting case study analyses of smart cities. Therefore, the main

© Springer Japan 2016

N. Tokoro, *The Smart City and the Co-creation of Value*,
SpringerBriefs in Business, DOI 10.1007/978-4-431-55846-0_3

task in the following chapters will be to elucidate in a theoretical framework consisting of the three components of "ba," "emergence" and "synthesis" the process of co-creation of value among the active agents participating in the construction of smart cities. In this process, "ba" will be the key component of the framework. "Ba" is the component that forms the foundation in a relationship-based strategy, and the process of co-creation of value is significantly influenced by the kind of "ba" that is formed.

Therefore, analysis of the "ba" should take priority above all else in the case analysis. In other words, the analysis of the nature of the "ba" must be appropriate and take into consideration factors such as circumstances that led to the formation of the "ba," the kind of relationships shared by the members participating in the "ba," and the manner in which the "ba" is being managed. Kanter (Kanter 1994) argues that whether collaboration among organizations is successful or not will depend on whether or not the organizations can develop a relationship of trust. Carley and Christie (1992) also indicate that collaboration among organizations in the early stages requires significant effort, and that the nature of the "ba" will have a decisively significant impact on value creation thereafter.

Because "emergence" and "synthesis," the other two components in the relationship-based strategy, are governed by the nature of the "ba," the discussion here will focus on the causal relationship of these components and "ba." Generally, when a "ba" has a high level of freedom with few restrictions and constraints from the external environment or specific active agents, bottom-up type "emergence" is more likely to occur. Moreover, it is believed that if a "ba" is managed appropriately, essential elements that lead to value creation will be selected from among the various "emergences," and this will lead to value creation through "synthesis." These views that have been justified in a general sense must be verified through case analyses, however. In the Yokohama Smart Community, "ba" with a high level of freedom were established, and the focus of the analysis here will be on aspects such as the kind of causal relationship the characteristics of such "ba" have with "emergence" and "synthesis" and, in that context, the kind of role leaders of such "ba" play.

Furthermore, the kind of competitive advantage acquired by participating companies through the Yokohama Smart Community initiative will also be another important focus of analysis. In a relationship-based strategy, it is believed that two competitive advantages can be acquired through co-creation among participating companies. The first is the competitive advantage derived from the integration of value created in the process of co-creation with a company's existing corporate value, and the second is the competitive advantage derived from the accumulation of tacit knowledge within a company through the exchange and integration of knowledge and know-how among companies during co-creation. The question one might ask, however, is whether these assumed competitive advantages in a relationship-based strategy are in fact ultimately manifested. This aspect must also be verified through the case analysis.

It has been indicated by many researchers to date that corporate partnerships and collaboration are beneficial in the establishment of a company's competitive advantage (Drucker 1995; Inkpen 1996; Doz and Hamel 1998). These studies, however, did not to validate the relationship of corporate partnerships and collaborations among companies with competitive advantages based on a certain theoretical framework. Instead, this chapter aims to establish the theoretical framework of relationship-based strategies and by verifying the relationship of co-creation among companies with competitive advantage under this framework to offer some insight on this subject.

3.3 Case Study: Yokohama Smart Community

3.3.1 Overview of Yokohama Smart Community

Launched in June 2011 as the initiative of a consortium comprised of companies, universities and local government, the Yokohama Smart Community is based on the principle of "realizing a city that supports everyday life and culture through science and technology as by learning from and utilizing nature." The participating members consist of 87 companies, representing a broad cross-section of industries including electronics, machinery, construction, energy, chemicals and housing. Also participating are researchers from five universities and the local government body of the city of Yokohama as a supporting organization.[1]

The Yokohama Smart Community has adopted a policy of leaving management of the consortium to the autonomy of the participating members. When participating members join the consortium, there is no need for them to pay any joining fees, nor is there any need to pay annual membership fees to cover administration of the consortium. The consortium operates on a system whereby the participating members "work without pay" and settle expenses for each project on their own. Of course, members are also free to withdraw from the consortium. Furthermore, the consortium owns no intellectual property rights, and it is understood that any intellectual property rights including technology and know-how developed in the course of a project belong to the participating members. Although the city of Yokohama, which is the local government, also belongs to the consortium, as a local government body it provides no financial support and participates solely in an advisory capacity. However, in the administration of the consortium, it would be inaccurate to say that no leaders exist. The role of "organizer" for initiating projects and coordinating matters among participating members does exist. In the case of the Yokohama Smart Community, the role of "organizer" is being fulfilled by two companies, dSPACE Japan and Smart Energy Laboratory, and these two companies serve as the representative and deputy representative of the Yokohama Smart Community respectively.

---

[1]Information concerning the Yokohama Smart Community is based on information provided by dSPACE Japan and various companies in attendance at the Smart City Week seminar held in Yokohama in October 2013.

In this way, the Yokohama Smart Community has formed a consortium as a "ba" with a high degree of freedom and high regard for the independence and autonomy of the participating members. The question one may ask, however, is what is the force that draws participating members to this consortium. On this point, the secretariat of the Yokohama Smart Community emphasizes two reasons: the first is empathy for a common vision, and the second is the existence of opportunities for validating technology. As stated earlier, the Yokohama Smart Community embraces the idea "to realize a city that supports life and culture through science and technology while learning from nature and utilizing nature." To explain in a little more detail the construct of this principle, the consortium uses plant cells as a metaphor. Inside plant cells are chloroplasts that absorb light from the sun and generate energy through photosynthesis, vacuoles that "store" this energy, and mitochondria that change it into supplies necessary for vital activities. Through these autonomous exchanges of energy, vital activities are maintained. Applying this concept of the functions of cells—production, storage, and consumption—as the smallest units of a plant to a social system is essentially the vision of the Yokohama Smart Community.[2] For example, by considering the houses people live in as plant cells, a power grid system as leaf veins, and the local communities as leaves, the objective of the Yokohama Smart Community is to construct in communities autonomous distributed energy networks based on a concept of life found in nature (Fig. 3.1).

Members participating in the consortium identify with the vision espoused by the Yokohama Smart Community and voluntarily join the consortium. In fact, the vision described above was conceived of by the two companies dSPACE Japan and Smart Energy Laboratory, and the consortium was formed by gathering members across a broad spectrum that shared an affinity with this vision. In other words, in the Yokohama Smart Community, the companies playing a leading role in the consortium articulated a vision and recruited members empathetic to this vision. In this way, they paved the way for sharing their vision among members. Existing research that analyzes collaborations and alliances among organizations points to the importance of establishing common knowledge among organizations (Cramton 2001; Clark 1996; Krauss and Fussell 1990). Therefore, it can be said that "sharing a vision" among members at an early stage is an important step in promoting smooth collaboration.

In "opportunities for validating technology," which is one of attractions of the consortium, members have opportunities to validate proprietary technology of their companies through projects by participating in the consortium. In other words, members can obtain data through practical application regarding how their companies' technology can be used and the potential it may have in the construction of autonomous distributed energy networks in communities as described above. In this way, they have the opportunity to accumulate valuable experience. Furthermore,

[2]This concept shares characteristics common to natural capitalism proposed by scholars like Hawken, P. and Lovins, A.

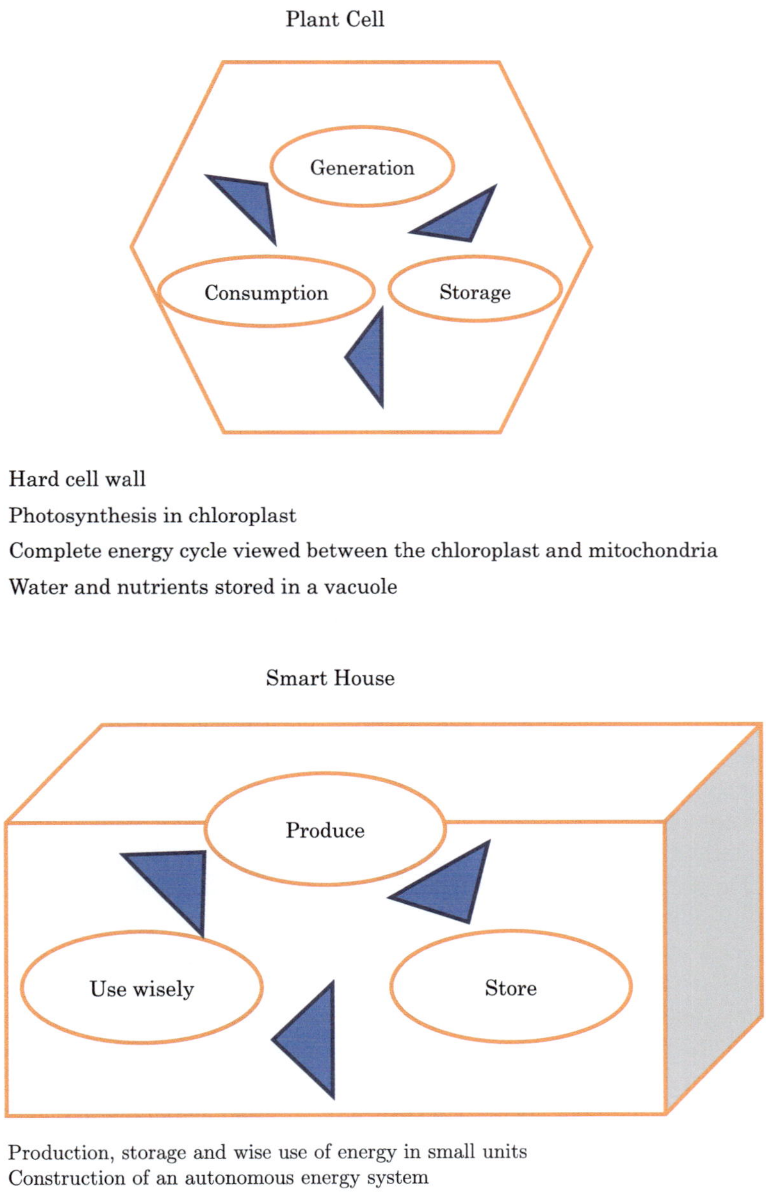

Fig. 3.1 Concept based on a viewpoint of life in nature. *Source* Prepared by the author after partially revising information provided by dSPACE Japan

various companies hailing from different industries are participating in the consortium, and it can be assumed that novel technological innovations can be achieved through the mutual resonance of proprietary technologies of diverse participating companies. It can also be said that the very value derived from such co-creation attracts participating members. Teece (2007) argues that the greater a company's dynamic capability is, the greater the possibility it has in generating innovation through co-creation with other companies. Therefore, the participation of companies with outstanding capabilities leads to further enhancing the attraction of the consortium.

3.3.2 Yokohama Smart Cell Project

As one of its activities, the Yokohama Smart Community has been conducting the "Yokohama Smart Cell" project. In this project, companies participating in the Yokohama Smart Community consortium constructed a model house called the "Smart Cell" to achieve the objective of eventually constructing in the community autonomous distributed energy networks learned from a viewpoint of life found in nature. As noted earlier, however, the management of the consortium in the Yokohama Smart Community is left to the autonomy of the members, and participation in the said project is voluntary. In fact, 17 companies are participating in this project.

In terms of the project overview, the plan was to build a two-story house called a "Smart Cell" on a block of land 330 m^2 in size within a housing exhibition park in Nishi-ku in the city of Yokohama. This model house is based on a view of life learned in nature as described earlier and replicates the functions of the vital activities of plant cells—production, storage, consumption. It does this through the autonomous exchange of energy and the functions of generating, storing and wisely using energy. Three core technologies used in the construction of the Smart Cell: passive technology, active technology and model-based development.

Passive energy refers to the method of controlling energy required for maintaining livability through methods that take advantage of nature rather than rely on power, such as insulation and natural ventilation. In the Smart Cell, passive technology is used in insulation material, high-fluidity concrete, water-permeable pavement, sunlight illumination, high-performance glass and super-insulated doors. Active technology, on the other hand, refers to an approach to achieving functions through lower energy consumption compared with energy consumed in the past for maintaining the building environment such as lighting, temperature and humidity as well as various living activities. Also included in this approach are methods for controlling energy consumption by flexibly accommodating energy through the creation of high-efficiency systems and energy-efficient equipment and the generation of energy. As new technologies through these methods, the Smart Cell makes use of smart distribution boards, optical heating and cooling systems, radiators, EV recharging stands and solar cells, among others. Model-based development refers to development methods used in the development of systems where safety and security are required despite their complex nature, such as the development of cars and aircraft, for example. In the Smart Cell, model-based development is used for the smart energy system and the development and verification platform (Fig. 3.2).

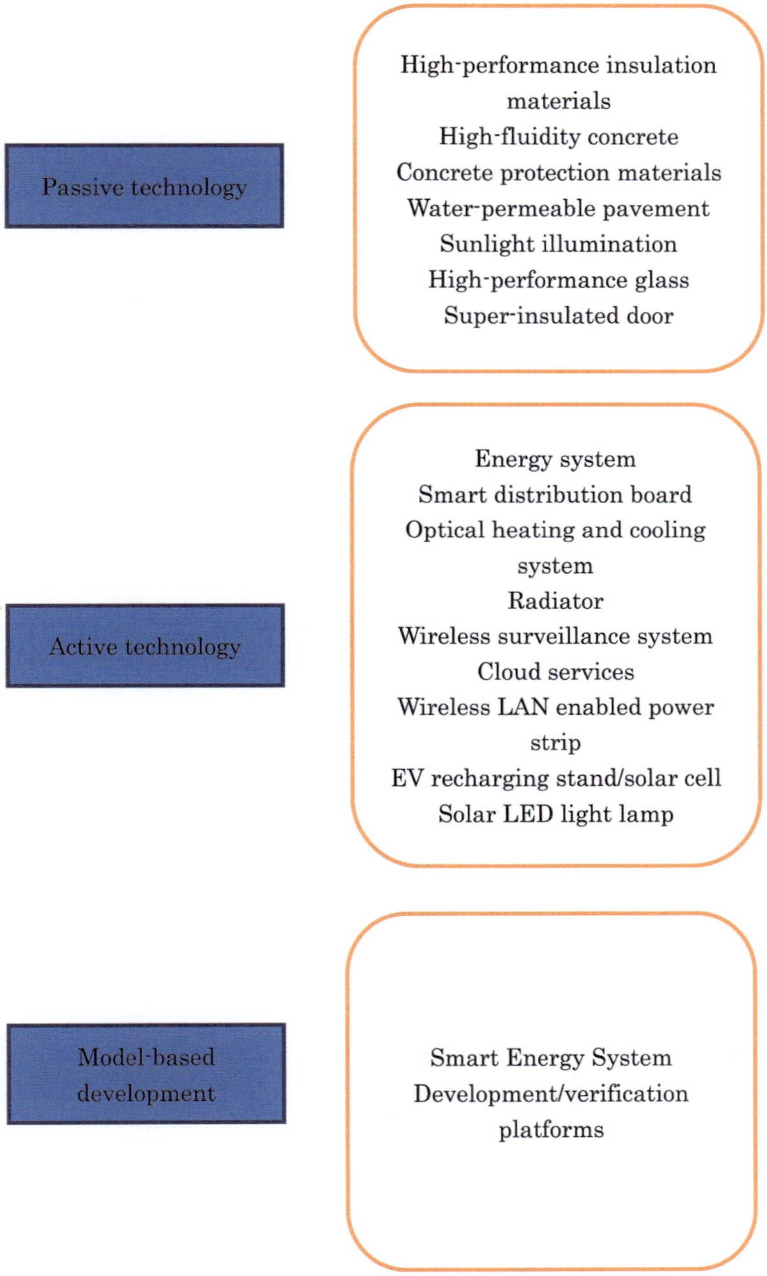

Fig. 3.2 Technologies Used in the Yokohama Smart House. *Source* Prepared by the author based on information obtained from the Yokohama Smart Community (2013) and the Yokohama Smart Community Seminar, Koubunn-Sha

In the Smart Cell project, the participating companies bring their respective strengths in technology to engage in the construction of Smart Cell. This chapter will next take a look at three companies participating in the Smart Cell Project and present a brief summary of their initiatives in the project. The companies are BASF Japan Ltd. (passive technology), Murata Manufacturing Company, Ltd. (active technology) and dSPACE Japan K.K. (model-based development).

3.3.3 Initiatives of BASF Japan

BASF Japan, a manufacturer of chemicals, is participating in the construction of Smart Cell in the area of passive technology in high-efficiency insulation materials, high-fluidity concrete and permeable pavement. Neopor, a high-performance thermal insulation material developed by BASF Japan, improves infrared light absorption reflective performance by approximately 20 %. This is achieved by adding black lead to existing thermal insulation material. As this material has a low water absorption rate, it can achieve sustainable thermal insulation performance for the serviceable life of the building. In the Smart Cell, Neopor is used as the thermal insulation material for the roof, walls, and foundation.

In addition, Smart Dynamic Concrete, a high-fluidity concrete also developed by BASF Japan, is also being used in the Smart Cell for the concrete section of the foundation. Smart Dynamic Concrete is capable of achieving superior fluidity and separating resistance using less cement than traditional ready-mixed concrete. Moreover, its low viscosity contributes to superior work performance and enables the construction of dense building structures high in workability. Elastopave®, a pavement material with enhanced water permeability, is used as the pavement for outdoor facilities. This is also a material developed by BASF Japan and is capable of achieving a high-strength pavement surface having open pore through the combination of polyurethane and mineral aggregates. One characteristic of water-permeable pavement using Elastopave® is the surface remains dry, so there is little danger of slipping. Moreover, rainwater does not splash off the surface, so there is no need for drainage piping or drains, or to provide for a catchment. Protectosil CIT, a material for preventing rebar corrosion, is another technology of BASF Japan used in the Smart Cell. This technology boasts efficacy in the two areas of water absorption prevention and rebar corrosion prevention, and is effective as a measure for prolonging the serviceable life of reinforced concrete structures. Protectosil CIT is already being used in structures in Japan in the Kaminoseki Bridge in Yamaguchi Prefecture and National Route 199 in Fukuoka Prefecture, for example. Overseas it has been used in the United States at the Ala Moana Pacific Center in Hawaii and in Denmark in the National Parliament.

By using the advanced technology it has developed as a chemical manufacturer and materials developed based on these in the construction of the Smart Cell in this manner, BASF Japan is attempting to verify their efficacy. As stated earlier, the areas BASF Japan is responsible for are the insulation, foundation, framework, and outdoor paving, and the technology relating to these areas is called passive technology. Other companies responsible for passive technology in the construction of the Smart Cell include Inosho K.K. (solar lighting systems technology), Nippon Sheet Glass Environment Amenity Co., Ltd. (high-performance glass technology) and Gadelius Industry K.K. (thermal insulation door technology), and the technology of BASF Japan must be reconciled with the technologies provided by these companies. This is essential in preventing any potential glitches among the technologies the respective companies bring to the Smart Cell. Without the reconciliation of technologies, it will be impossible to demonstrate the overall effects of the Smart Cell. For example, Neopor, a high-performance thermal insulation material developed by BASF Japan, is used in the Smart Cell roof and walls but unless performance adjustments are made between this thermal insulation material and the solar lighting system provided by Inosho, the thermal insulation effects of the Smart Cell will be compromised. Moreover, a new indoor environmental adjustment system called the "optical heating and cooling system" developed by the AnnyGroup has been adopted in the Smart Cell as active technology. This system is designed to maintain the overall interior space at a comfortable temperature by controlling the indoor sensory temperature more efficiently. This is achieved through the resonance of far infrared generated from the interior walls, which are made of a special ceramic, the interior material coating on the ceilings, and optical cooling and heating radiators. Here too, reconciliation between the technologies of BASF Japan's high-performance insulation material and the optical heating and cooling system was carried out, and the resonance achieved through reconciliation of these two technologies made it possible to achieve high performance in the Smart Cell as a whole.

To promote smooth reconciliation in technologies between companies of different industrial sectors in this way, high-quality dialogue and the development of relationships of trust on the part of the persons in charge are vital. Kodama argues that companies of different industries in particular require management that has an adequate understanding of each other's technologies and is capable of making appropriate judgments of their synergistic effects (Kodama 2007, 2010). In such situations, companies must also consider how they will protect their companies' intellectual assets (Teece 2000). In "ba" of this nature, decisions are made in a mixed environment of co-creation and competition.

3.3.4 Initiatives of Murata Manufacturing Company, Ltd.

Manufacturer of electronic materials and components, Murata Manufacturing Co., Ltd. is participating in the construction of the Smart Cell in the area of creating an energy system for next-generation smart houses. In the construction of the Smart Cell, this area comes under "active technology." The Smart Cell requires the functions of producing, storing and wisely using energy, and it can be said that the development of a system that can control such functions is technology at the very core of the Smart Cell. Making use of technology it acquired over the years through research and development and the production of functional ceramic-based electronic devices, Murata Manufacturing has taken up the challenge to construct an energy system for the Smart Cell, a new area for the company. To create a system to provide for the efficient circulation of energy in the Smart Cell, the combination of broad-ranging technologies was essential. In a product that Murata Manufacturing refers to as the "demonstration experiment device" for the energy system of next-generation smart houses, the company has incorporated the following functions:

- Capability to determine at all times the electric power generation capacity of solar cells, the residual amount of energy in storage cells, the amount of electric power used in the Smart Cell, and to conduct the optimal allocation of energy
- Autonomous operation during power outages and back up for grid power through peak cuts and power shifts

In addition, Murata Manufacturing is also said to be considering use of this device with cloud data for weather forecasting and electricity forecasting, and energy usage linking smart houses and communities.

The Yokohama Smart Community vision is to create "an energy system learned from nature" and it goes without saying that use of natural energy including photovoltaic power generation will play a central role in the Smart Cell. At the same time, however, there are some concerns about natural energy in terms of stable electric power supply, and coordination of natural energy with grid power will be essential. Murata Manufacturing's demonstration experiment device has the ability to determine in real time the amount of electric power generated from natural energy, the amount of power stored, and the amount of electricity used. It also has the ability to enable the grid system to increase or decrease electricity supply according to circumstances.

This demonstration experiment device is core technology of the Smart Cell, and therefore, as a matter of course, must be reconciled with other technologies. For example, reconciliation with the Smart Distribution Board developed by Kawamura Electric Inc. was essential in the coordination of the energy system between natural energy and grid power. Connection of both sources optimizes energy use through storage and the implementation of peak cuts and peak shifts. Furthermore, the optical heating and cooling system developed by the AnnyGroup mentioned earlier has also been incorporated into the Smart Cell energy system, and it is only by linking this to the energy system that the system can demonstrate efficacy. Naturally, coherence and reconciliation were carried out to achieve this. In the same way, coordination and reconciliation were carried out between technologies such as the Cloud Service provided by Ubiquitous Corporation and the EV Charger, solar cells and solar LED lighting provided by Star Engineering Co., Ltd.

Understandably, Murata Manufacturing is not singlehandedly responsible for the construction of Smart Cell energy system. Moreover, because the above technology is core technology of the Smart Cell, the two companies in leadership positions in the Yokohama Smart Community consortium, that is, dSPACE Japan and Smart

Energy Lab, are also involved. In fact, the overall system design of Smart Cell energy system was done by Smart Energy Lab. The originator of the concept of building an "energy system learned from nature" was Smart Energy Laboratory, which has also been responsible for the basic design of the Smart Cell in the Smart Cell Project. Smart Energy Laboratory's role in the construction of the energy system has been consulting in matters ranging from concept making and development to verification. dSPACE Japan, which has played a leading role along with Smart Energy Laboratory in the project, is involved in the methods for model-based development discussed later and the provision of various types of equipment required in the construction of the energy system. In addition, Murata Manufacturing has been involved in the construction of the energy system through development of equipment for demonstration experiments. Essentially, it can be said that the Smart Cell energy system is a product of the co-creation of three companies: Smart Energy Lab, dSPACE Japan, and Murata Manufacturing.

While it is true that an energy system is an information management system using IT, whether a company views it simply as an information management system or decides to delve into attendant underlying issues of "knowledge" and make efforts to further exploit this knowledge will change the significance of the information management system in business management. There are said to be many companies that mix information management and knowledge management (Nonaka and Takeuchi 1995; Nonaka et al. 2008) but whether they view this as an issue of superficial information management or as an issue of deep knowledge that could be exploited will change the role of management of the companies involved.

3.3.5 *Initiatives of dSPACE Japan*

As a company, dSPACE Japan is involved in the development of mechatronics control systems, and this technology is widely used in areas of forefront technology for automobiles and aircraft, among others. In the construction of the Smart Cell, dSPACE Japan is responsible for model-based development. Model-based development is a method for providing a new environment for the development of electric power resources with outstanding efficiency and traceability in the field of energy system development where solar cells, storage cells, and grid power are integrated. In reality, the development of a system with more advanced, complex functions and strict safety standards is not an easy feat in product development. There is also the issue of costs required for such development. In model-based development, dSPACE Japan adopted a groundbreaking development method by introducing graphically represented mathematical models and creating control logic virtually. This makes the simulation, automation, and reproducibility of an energy system possible.

However, dSPACE Japan is not singlehandedly responsible for model-based development, which involves co-creation with other companies in areas such as the design and validation of system specifications and controls. For example, in areas involving energy system development platforms, simulation environments and prototype development environments that utilize numerical calculation tools and electric power simulators, dSPACE is engaged in co-creation of technology with Smart Energy Laboratory. Moreover, the company iTest is also involved in areas such as module verification, interfacing verification and conformity measurement.

In model-based development too, high-quality dialogue and the development of relationships of trust on the part of the persons in charge are essential in promoting smooth co-creation among companies. At the same time, however, each company must be mindful of protecting its intellectual assets. In other words, as stated earlier, decisions in such "ba" are made in a mixed environment of co-creation and competition. Although the objective in the Smart Cell Project was to construct an energy system for a single home, the ultimate goal of the Yokohama Smart Community is to construct the same kind of energy system for an entire community and to create autonomous distributed energy networks. Therefore, the Smart Cell is just one milestone in achieving this. dSPACE Japan, which is in a leadership position in the Yokohama Smart Community, is therefore required to fulfill the role of strategy innovator (Hamel 1998) to create a grand design for constructing a system for an entire community based on knowledge obtained from this project.

3.4 Fukuoka Smart House Consortium and Nagasaki Smart Society

As stated earlier, the Yokohama Smart Community was inaugurated in June 2011 but it has close ties with the Fukuoka Smart House Consortium established in Fukuoka City the previous year in June 2010. Established by 13 organizations including Smart Energy Lab, dSPACE Japan, Texas Instruments Japan and Sojo University, the Fukuoka Smart House Consortium had 68 participating organizations as of September 2014. The objective of this consortium was to learn from the behavior of plant cells in the natural world that autonomously control energy to create a sustainable energy system for "producing, storing, and wisely using energy." The consortium constructed and conducted demonstration experiments on a two-story smart house on land in Fukuoka Island City in Higashi-ku in Fukuoka City. Based on this concept of life learned from nature, the consortium conducted demonstration experiments on the following:

- Smooth introduction of natural energy
- Consistent control of grid power
- EV (electric vehicle) charging at home
- Autonomous energy control during power outages
- Exchange of energy in both directions
- Advanced integration of energy and information
- Introduction of model-based development methods

- Development environment where internal performance and the external environment can be validated
- Linkage with meteorological data

The companies participating in the consortium brought their respective technologies and products to conduct demonstration experiments at the Fukuoka Smart House project. Honda Sol Tec (HST) was in charge of the photovoltaic power generation system, Zephyr the small-scale wind power generation system, Baysun the lithium-ion batteries and Aval Nagasaki Corporation the electric power control system. Using SCALE, a power supply circuit simulator tool, the Electronics Research Lab of Sojo University assumed the role of conducting simulations of the flow of electricity in the smart house as a whole as well as among the component parts and conducted comparative investigations of the results with results of demonstration experiments.

The rule that participating companies take part in demonstration experiments without pay is a characteristic of this consortium. The consortium has neither joining fees nor budget. Nor has it any specific plans for projects. What it does have is a shared vision to create a sustainable energy system for "producing, storing, and wisely using energy" learned from the behavior of plant cells in the natural world, which autonomously control energy. Therefore, in promoting the project, those in charge of various areas had to work out their own plans. As a result, a "ba" for discussing plans among the parties in charge naturally came into being.

Moreover, the consortium does not depend on public subsidies and works on the principle of autonomous management. In cases where management of a project depends on public subsidies, there is a likelihood that activities will stall if subsidies are cut, and consortium activities including management of its budget and outcomes will be restricted. The local government of Fukuoka City has provided a certain level of support for the project such as providing on loan a brick house that the city owns and including the project in the Island City Project within the Green Asia International Strategic Comprehensive Special Zone. However, the local government's support has been entirely in an auxiliary capacity and the activities of the Fukuoka Smart House Consortium are extremely autonomous, consisting mainly of companies and organizations that share the vision of the consortium, who meet voluntarily and provide their respective technologies and products at their own expense.

In fact, the activities of the Fukuoka Smart House Consortium led to the creation of the Yokohama Smart Community. It began when the local government of Yokohama, which was attracted to the activities of the Fukuoka Smart House Consortium, made a request to Smart Energy Lab and its joint partner dSPACE Japan, which were in charge of the consortium concept making, to undertake the same kind of activities in Yokohama. Therefore, at the request of the city of Yokohama, the Yokohama Smart Community was inaugurated in June 2011, one year after the establishment of the Fukuoka Smart House Consortium, with the objective of undertaking in Yokohama the activities that were being conducted in

Fukuoka. The vision espoused by the consortium and the management of the consortium were to be the same as in Fukuoka. Moreover, the core members—Smart Energy Lab, dSPACE Japan, Murata Manufacturing, Ubiquitous and Sojo University were the same members. While the demonstration experiments of the technologies for the smart house featured prominently in the Fukuoka Smart House Consortium, the objective of the activities in the Yokohama Smart Community was not only to conduct demonstration experiments on the technology alone but also to create a community model that would foster the development of nature, art and culture. In other words, while following approaches of Fukuoka along the basic lines of vision and consortium management methods, the Yokohama Smart Community aimed to expand the scope of activities and enhance the project as a whole.

In addition, the circle of such consortium activities appears to be broadening. In July 2012, the Nagasaki Smart Society, which was to become the third consortium, was established. Nagasaki Prefecture is located at the tip of western Japan, and 45 % of its prefectural land consists of isolated islands. It also has a coastline of 4000 km. With remote islands richly endowed in natural beauty such as Goto, Iki and Tsushima as well as plentiful forests and maritime resources, Nagasaki is considered to be a region with high environmental value. Following on from initiatives in Fukuoka and Yokohama, a consortium was formed in Nagasaki with the objective of realizing compact, smart communities that capitalize on the strengths of the region. The basic tenets of this consortium, including the vision and the management of the consortium, are similar to those of the Fukuoka and Yokohama consortia. Moreover, the core members are also the same. Specific activities of the consortium include the holding of seminars and initiatives in human resource training, which have already commenced. At the same time, the consortium cooperates with the Huis Ten Bosch project established 20 years ago, which is based on the concept of comprehensive symbiosis with nature of a town that has continued to exist for 1000 years.

In this way, the activities that commenced with the Fukuoka Smart House Consortium have subsequently spread to the Yokohama Smart Community and the Nagasaki Smart Society. As stated earlier, consistent principles exist in the pattern of these projects. Among the common aspects shared by the three consortia are the vision and consortium management methods embraced by each consortium. The vision, as explained earlier, is to build autonomous distributed energy networks in the community that imitate the functions of plant cells by "producing, storing and wisely using energy" based on a view of life learned from nature. Another common aspect of the consortia is members have no obligation to pay joining fees or annual fees, and consortia management is based on the volunteer service of the participating members. Moreover, the core members are the same. On the other hands, the activities of each consortium have their own unique characteristics. For example, while conducting demonstration experiments of individual smart house technologies is the main objective of the consortium project in Fukuoka, in the case of Yokohama, in addition to conducting demonstration experiments, the consortium aims to create a community where nature, art and culture are integrated. In the case of Nagasaki, there is a strong consciousness of achieving local production and local consumption of energy. This consortium also has its sights set on power generation harnessing Nagasaki's rich natural resources mentioned earlier, particularly the geothermal heat of the Shimabara Peninsula and heat from hot springs. Moreover, while there is no difference in the core members of the respective consortia, the participation of local companies in Fukuoka, Yokohama and Nagasaki add a characteristic flavor to the consortia.

Activities of this nature are also expected to increase in regions throughout Japan in the future. Already Okinawa has been named as the next candidate area. By providing a free, open place to participating companies, these consortium activities enable participating companies to amass valuable intellectual resources within their organizations. Intellectual resources always exist within an organization (Cohen and Prusak 2000) but how they are regarded and how they are utilized will significantly affect the status of a company. This aspect is as explained earlier.

3.5 Implications

To conclude this chapter, I would like to summarize implications observed from the above case analysis. Considering a relationship-based strategy as a theoretical framework, this chapter analyzed initiatives of the Yokohama Smart Community. In doing so, it focused on the following three points and questions as the focus of analysis.

(1) Analysis of "ba"
What were the circumstances under which "ba" were formed? What was the relationship of the members participating in the "ba?" How were the "ba" managed?
(2) Analysis of "emergence" and "synthesis"
What kind of impact did the nature of the "ba" have on "emergence" and "synthesis?"
(3) Analysis of the competitive advantages gained by the participating companies
What kind of competitive advantages were the companies that participated in the consortium able to gain?

First, in regard to analysis of the "ba" in (1), as already mentioned a number of times, the "ba" formed in the Yokohama Smart Community was a free, open place where the self-motivation and autonomy of participating members were respected. Underlying the formation of this "ba" were the unique circumstances of the participating members who were united only through a "shared vision." In other words, the Yokohama Smart Community was not an arrangement where a certain leader company existed and under whose strong leadership the consortium was formed. Nor was it a case where other members participated in the consortium because of various relationships of interest shared with the leader company. Furthermore, it was not a consortium initiated by a government body which recruited consortium members by offering subsidies or other incentives. The consortium was formed by various organizations that shared a vision to create an energy system learned from “a view of life in nature” and which came together at their own initiative.

This self-motivated “ba” placed almost no restrictions on participating members. To participate in the consortium, there were no joining fees nor was there any obligation to pay membership dues for administration. Moreover, the consortium had neither budget nor plans. At their own initiative, participating members voluntarily discussed among themselves administration policies of the consortium and made decisions on the implementation of projects. The management of a project entailed having the participating members bring their respective technologies, and expenses required for implementing the project were settled upon its completion. Therefore, relationships between members of the Yokohama Smart Community were extremely flat. In other words, relationships between members were not governed by superior-subordinate relationships based on a capital relationship, or relationships of interest based on business affiliations and, therefore, allowed for free, open discussion.

Furthermore, in regard to management of “ba,” the existence of a highly autonomous “ba” also conversely leads to the absence of appropriate “ba” management. In other words, the participating members’ emphasis on autonomy and independence makes implementation of management that might obstruct these difficult. Consequently, there is a risk that this stance could ultimately lead to a laissez-faire policy of no management. In response to the question as to whether or not appropriate “ba” management is being conducted in the Yokohama Smart Community, it would not be possible to give a clear answer at this stage. However, it is clear that there is no evidence management required for revitalizing the “ba” and bringing about value creation is being performed by a specific party as Itami (Itami 1999) describes. As stated earlier, in the Yokohama Smart Community there are companies who assume a leading role in the consortium. They are the two companies Smart Energy Laboratory and dSPACE Japan. However, these companies do not demonstrate leadership in the conventional sense by playing a dominating role at the “ba.” In fact, Yoshimichi Nakamura, CTO and founder of Smart Energy Laboratory, who founded the consortium and is also originator of the consortium vision, plays an instrumental role in the consortium. A theoretical support pillar of the consortium, he is considered to be a person capable of initiating the kind of innovation elucidated in the research of Tomala and Senechal (Tomala and Senechal 2004), but his capability in “ba” management is an unknown quantity.

Next, in regard to analysis of “emergence” and “synthesis” in (2), these are closely related to the nature of the “ba.” As stated earlier, it is believed that there is a positive correlation between a “ba” with a high degree of freedom and “emergence.” Moreover, it is also believed that value creation through synthesis will occur when the "ba" is appropriately guided. The question remains, however, whether this will occur as expected. The "ba" formed in the Yokohama Smart Community is a "ba" with a high degree of freedom. As a result, quite notable "emergence" occurred. The circle of activities that started with the Fukuoka Smart House Consortium expanded to the Yokohama Smart Community and on to the Nagasaki Smart Society. While the core of the respective consortia remained solid, the consortia have continued to evolve with the addition of new members. This series of developments had not been planned beforehand. In other words, the formation of consortia in Fukuoka, Yokohama and Nagasaki in succession to carry out projects in those locations was not part of a plan conceived in a top-down manner. It was in the course of executing other projects that these ideas arose spontaneously. The Fukuoka and Yokohama consortia are said to have increased their membership by holding regular seminars where prospective members hoping to participate come into contact with the vision espoused by the consortium. Moreover, among the participating companies there are said to be many companies that were initially unable to obtain approval from top management to participate in the consortia because projects of the consortia had neither budgets nor concrete plans. It was supposedly mid-level engineers who persuaded management to come on board. They not only understood the significance of the project but also as engineers wanted to confirm the potential of their companies' proprietary technology. As a result, the consortia have attracted various highly motivated people.

Various studies on emergence to date (Bower, Gilbert 2007; Mintzberg 1973, 1978, 1990; Mintzberg and Waters 1985; Quinn, 1978, 1980; Burgelman, 1983, 1994, 2002) point to the importance of freedom within the organization, and ideas springing up from middle management level in the workplace. The case of the Yokohama Smart Community already adequately confirms such findings.

On the other hand, it is difficult to obtain adequate insights from case analyses regarding synthesis as of the present. As indicated earlier, even if various incidences of synthesis occur through "ba," they do not necessarily immediately result in value creation. Synthesis is a process where elements created through "emergence" are first scrutinized and then recombined to create new value. In the Fukuoka and Yokohama consortia, some noteworthy technical innovations are said to have occurred through demonstration experiments in the smart house and Smart Cell, and it is certain that there are indications these will result in value creation. However, the presence of management that could possibly bring about value creation is weak. As a result, the overall prospects for synthesis remain unclear. Nevertheless, as the Fukuoka Smart House Consortium, Yokohama Smart Community, and Nagasaki Smart Society demonstrate a broadening in their ring of activities, their steadfast sharing of the vision to build energy networks in communities based on "a concept of life learned from nature" is a special characteristic of these consortia. Therefore, if management that links this vision to value creation is exercised with certainty, it will be possible to validate value creation through synthesis.

Finally is the analysis of acquisition of a competitive advantage of the participating companies in (3). Earlier it was stated that companies participating in the consortium could be assumed to acquire two competitive advantages. The first was the competitive advantage gained through the integration of value created through co-creation with a company's existing corporate value, and the second was the competitive advantage achieved through the accumulation of tacit knowledge within the company resulting from the exchange and integration of knowledge and know-how between companies in the process of co-creation. Of the two, at this stage it is not possible to confirm the first. When cooperation among companies in a consortium type of "ba" is seen to create new value and that value is integrated with a corporate value and clearly enhances corporate value, the participating companies can be deemed to have gained a new competitive advantage. Under the present circumstances, however, it is not possible to make such a judgment. In this regard, as expected, the presence of management that assumes leadership of the "ba" and achieves value creation from cooperation among companies is significant. In the case of the Yokohama Smart Community, "ba" management is weak. Consequently, there are difficulties in achieving external recognition of value creation taking place in the consortium.

On the other hand, the second competitive advantage can be confirmed to some extent. Demonstration experiments of technologies in the smart house and Smart Cell has been the main objective of the Fukuoka and Yokohama consortia and the activities of these consortia, which are achieving one technological innovation after another, are attracting attention both in Japan and overseas. Among the technologies being verified through these consortia are development kits based on model-based development methods, power supply circuit simulators using energy control technology, energy control managers that store in condensers and release from condensers PV electricity, and that mix and use AC-DC conversion and system power and special processors for energy control. These achievements are results of co-creation among companies in the consortia as "ba." Moreover, various data obtained in the course of co-creation accumulate within these companies and become valuable assets in the development of new products and systems. Not only that, it goes without saying that various types of knowledge and know-how obtained through co-creation, accumulate as tacit knowledge within a company.[3] However, Eisenhardt and Santos (2002) indicated that such tacit knowledge is difficult to perceive from outside a company and its validation is also difficult. Consequently, positive evidence of the causal relationship between knowledge creation and competitive advantage is scarce. Teece (1998) takes the view that in knowledge-based companies the core of a company's competitive advantage lies in its intangible assets, and this causal relationship becomes clear in the course of long-term qualitative analyses on a time axis (Fig. 3.3).

[3]There are two reasons why validating tacit knowledge is difficult. The first is knowledge of this nature constitutes corporate secrets that are highly confidential, and therefore companies do not release this information externally. The second reason is companies themselves lack an awareness of the existence of tacit knowledge within their organizations.

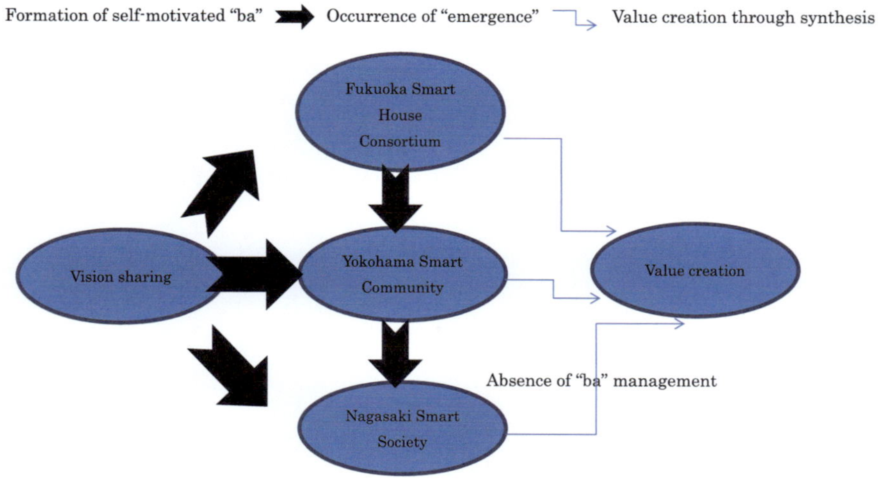

Fig. 3.3 Process of value creation in a consortium. *Source* Prepared by the author

References

Bower, J.L., & Gilbert, C.G. (eds.). (2007). *From Resource Allocation to Strategy*. Oxford: Oxford University Press.

Burgelman, R. A. (1983). A process model of internal corporate venturing in the diversified major firm. *Administrative Science Quarterly, 28*(2), 223–244.

Burgelman, R. A. (1994). Fading memories; A process theory of strategic business exit in dynamic environments. *Administrative Science Quarterly, 39*(1), 24–56.

Burgelman, R. A. (2002). Strategy as vector and the inertia of coevolutionary lock-in. *Administrative Science Quarterly, 47*(2), 325–357.

Carley, M., & Christie, I. (1992). *Managing Sustainable Development*. London: Earthscan.

Clark, H. (1996). *Using Language*. New York: Cambridge University Press.

Cohen, D., & Prusak, L. (2000). *In Good Company; How Social Capital Makes Organizations Work*. Boston, MA: Harvard Business School Press.

Cramton, C. D. (2001). The mutual knowledge problem and its consequences for dispersed collaboration. *Organization Science, 12*(3), 346–371.

Doz, Y. L., & Hamel, G. (1998). *Alliance Advantage; The Art of Creating Value through Partnering*. Boston, MA: Harvard Business School Press.

Drucker, P. F. (1995). The network society. *Wall Street Journal.* March 29, 12.

Eisenhardt, K. M., & Santos, F. M. (2002). Knowledge-based view; A new theory of strategy? In A. M. Pettigrew, H. Thomas, & R. Whittington (Eds.), *Handbook of Strategy and Management* (pp. 139–164). London: Sage.

Hamel, G. (1998). Strategy innovation and quest for value. *Sloan Management Review, 39*(2), 7–14.

Inkpen, A. C. (1996). Creating knowledge through collaboration. *California Management Review, 39*(1), 123–140.

Itami, H. (1999). *Management of "Ba"; A New Management Paradigm*. Tokyo: NTT Publishing.

Kanter, R. (1994). Collaborative advantage; Successful partnerships manage the relationship, not just the deal. *Harvard Business Review, 72*(4), 96–108.

Kodama, M. (2007). *Knowledge Innovation; Strategic Management as Practice*. Cheltenham: Edward Elgar Publishing.

Kodama, M. (2010). *Boundary Management; Developing Business Architectures for Innovation*. Heidelberg, Dortrecht, London, New York: Springer.

Krauss, R. M., & Fussell, S. R. (1990). Mutual knowledge and communicative effectiveness. In J. Galegher, R. E. Krant, C. Egido (Eds.), *Intellectual Teamwork; Social and Technical Bases of Collaborative Work*. Hillsdale, NJ: Erlbaum, 111–146.

Mintzberg, H. (1973). Strategy-making in three models. *California Management Review, 16*(2), 44–53.

Mintzberg, H. (1978). Patterns in strategy formulation. *Management Science, 24*(9), 934–948.

Mintzberg, H. (1990). The design school; Reconsidering the basic premises of strategic management. *Strategic Management Journal, 11*, 171–195.

Mintzberg, H., & Waters, J. A. (1985). Of strategies, deliberate and emergent. *Strategic Management Journal, 6*, 257–272.

Nonaka, I., & Takeuchi, H. (1995). *The Knowledge-Creating Company*. New York: Oxford University Press.

Nonaka, I., Toyama, R., & Hirata, T. (2008). *Managing Flow; A Process Theory of the Knowledge-Based Firm*. London: Palgrave macmillan.

Quinn, J. B. (1978). *Strategic Change; Logical Incrementalism* (pp. 7–21). Fall: Sloan Management Review.

Quinn, J. B. (1980). *Strategies for Change; Logical Incrementalism*. Englewood Cliffs, NJ: Irwin.

Teece, D. J. (1998). Capturing value from knowledge assets; The new economy markets for know-how and intangible assets. *California Management Review, 40*(3), 55–79.

Teece, D. J. (2000). *Managing Intellectual Capital; Organizational Strategic and Policy Dimensions*. New York: Oxford University Press.

Teece, D. J. (2007). Explicating dynamic capabilities; The nature and microfoundations of (sustainable) enterprise performance. *Strategic Management Journal, 28*(13), 1319–1350.

Tomala, F., & Senechal, O. (2004). Innovation management; A synthesis of academic and industrial points of view. *International Journal of Project Management, 22*(4), 281–287.

Yokohama Smart Community. (2013). *Yokohama Smart Community Seminar*. Tokyo: Koubun-sha.

Chapter 4
Co-creation of Value Through Initiative of a Leader Company and Collaboration of Participating Companies—Case Study of Fujisawa Sustainable Smart Town

4.1 Introduction

In this chapter, I will examine a smart city construction project which Panasonic, one of Japan's major electronics manufacturers, is currently promoting in Fujisawa City in Kanagawa Prefecture. This project is based on Panasonic's commitment under its new management strategy to provide "comprehensive solutions for the entire house, entire building, and entire town," and Panasonic is playing the leading role in all aspects of the project from the drafting of the plans to their execution. In addition to Panasonic, however, a large number of partner companies are participating in the project. Therefore, it cannot be said that it is a 100 % Panasonic project. Furthermore, the local government of Fujisawa City is also involved in the project to some extent.

In view of these factors, therefore, the analysis of the project must bear in mind the two aspects that the project has. The first is the existence of Panasonic as a strong leader that initiated the project, and the fact that the project will proceed at times in a top-down manner under Panasonic's leadership. The second is the fact that the project will proceed at times in a bottom-up manner through the collaboration of Panasonic and the partner companies. Determining what kind of value is created when these top-down and bottom-up approaches merge will be the focus here, and analyzing this phenomenon will be the central issue in this chapter.

4.2 Analytical Framework

As in Chap. 3, I would first like to confirm the analytical framework before proceeding with the case analysis. As I stated in the introduction, the case study to be taken up in this chapter is a project in which Panasonic plays the leading role and,

© Springer Japan 2016

N. Tokoro, *The Smart City and the Co-creation of Value*,
SpringerBriefs in Business, DOI 10.1007/978-4-431-55846-0_4

therefore, will strongly reflect the intentions and strategy of Panasonic. Furthermore, it is understood that other companies participating in the project will basically be required to conform to the vision and plans conceived of by Panasonic, and it can be reasonably expected that under such circumstances the nature of the "ba," or shared context in motion, where the participating companies converge will to some extent be restricted in autonomy and freedom. The focus of the analysis here will be to determine what kind of nature the "ba" will have and whether some kind of mechanism or procedure is in place to ensure the autonomy and independence of the participating members when a strong leader company exists. Furthermore, in regard to management of the "ba," while it can of course be assumed that the leader company will take control of management of the "ba," details of that management and the quality of management must also be analyzed. For example, if the leader company practices management that is biased toward top-down or heavy-handed leadership, there is a possibility the "ba" will atrophy, leading to a situation where it will be difficult for co-creation of value to occur. Determining whether the leader company is skillfully managing the "ba" in a manner that encourages value co-creation with the participating companies and, if so, what kind of methods it employs will also be the subject of this analysis.

The analysis of "emergence" in this case study will require more in-depth analysis. As I already stated earlier, it is generally known that the manifestation of "emergence" occurs in a bottom-up manner in a "ba" where there is a high degree of autonomy and freedom. The case presented here, however, will not necessarily conform to this general perception. In view of the nature of the project, a scenario similar to one of those suggested by proponents of strategic planning (Ansoff 1965; Andrews 1971; Steiner 1969) can be imagined. In such a scenario the leader drafts a plan and executes it in a top-down manner and the respective members are required to follow the plan with uniform behavior. In such an environment, the occurrence of "emergence" would be unlikely under normal circumstances. On the other hand, according to studies by Mintzberg (Mintzberg 1973, 1978, 1990), when emergence is discussed in a context where there is an assumption of an intended strategy of a top leader, and the participating members are granted a certain degree of autonomy, there is a possibility interesting emergences will occur in the course of merging top-down and bottom-up results.

This also applies to the analysis of synthesis. The act of extracting from emergence essential elements that will lead to value creation and recombining these is what is meant by "synthesis." In the case study here, it is assumed that value will be created in the course of merging "analysis" and "synthesis." Put another way, this is essentially a scenario where the process of intended value creation based on the intended strategy of the leader company and the process of unintended value creation that springs up at the workplace merge to create new value.

4.3 Case Study: Fujisawa Sustainable Smart Town (Fujisawa SST)

In this section I will conduct a case analysis of the Fujisawa Sustainable Smart Town (Fujisawa SST) project which Panasonic is promoting in Fujisawa City in Kanagawa Prefecture. Launched in 2011 and aiming for completion in 2018, this project is currently in progress. While Panasonic is largely responsible for the project, companies of different industries are also active participants, and it is with their cooperation that construction of the smart city is making progress. There are also expectations that the exchange and merging of technologies, knowledge and know-how that take place among the companies through this process will lead to the creation of new value.

4.3.1 Overview of Fujisawa SST

Located on a former Panasonic plant site (approximately 19 ha) in Fujisawa City in Kanagawa Prefecture, Fujisawa SST is a project that aims to create an innovative town of approximately 1000 households and 3000 residents. Total cost of the project will be approximately 60 billion yen. Construction, which began in 2011, is expected to be completed in 2018. Some specific details of the project are described below.[1]

First of all, in the Fujisawa SST, all districts of the town including all homes, facilities and public zones will be equipped with solar power generation systems and storage batteries for home use as standard fittings. In shops and stores, various devices for energy-creation, energy-saving, and energy-storage will also be introduced in the four areas of cooling, lighting, heating and water.

In public spaces, "eco cycle packages" will be introduced for the use of recharging facilities for electric cars and plug-in hybrid cars and facilities such as electric-assisted bicycles and solar parking areas. Also included in the project plans are mobility sharing services including eco cars and electric cars, security services with optimal control using a combination of lighting, sensors and monitoring cameras, and healthcare services that will provide facilities to enable the elderly to live in comfort.

As a community platform to support these systems, the project will also provide terminals and a one-stop portal that will offer applications for accessing all kinds of services. Homes will be linked by SEG (Smart Energy Gateway: a system for the uniform management of networked electrical appliances and equipment within the home) which will make energy "visible." Also a system that can be operated from

[1]Description of the project is based on materials provided by a representative of Panasonic during an interview conducted by the author on February 26, 2014 at the Panasonic Tokyo Shiodome Building as well as information made public on the company's website.

the comfort of one's living room will enable residents to manage flash sale notices of commercial facilities and reservations of facilities. By implementing such initiatives, Fujisawa SST aims to cut the town's overall carbon dioxide emissions by 70 % and reduce water consumption in everyday living by 30 % compared to 1990 levels.

As a project led by Panasonic, Fujisawa SST strongly reflects the intentions of Panasonic from the formulation of the vision to the drafting and implementation of plans. Essentially, the project has been implemented under Panasonic's strategy to provide "solutions for the entire home, entire facilities, and entire towns," and Panasonic hopes to use this strategy as a springboard for its own transformation from a producer of individual white goods and devices to date to a creator of new urban spaces for the 21st century. It is said that Panasonic intends to establish a business model based on knowledge gained through the Fujisawa SST and has aspirations to take on the world market in the future by making this new business one of its pillars. Therefore, other companies participating in the project apart from Panasonic are expected to understand this Panasonic strategy and to fulfill their respective roles in line with Panasonic's intentions. In addition to Panasonic, there are eight other companies participating in Fujisawa SST at the time of the project's launch in 2011, and these companies have been promoting the business plan. These eight companies hail from diverse business sectors and include a residential developer, real estate company, gas company, financial institution, and trading company. The division of responsibilities of the respective members in the project are roughly as shown in Table 4.1.

Table 4.1 Roles in Fujisawa SST

Accenture	• Smart town conceptualization, service model planning and promotion • Smart town platform support in light of global trends
Orix	• Service planning for increased overall town value, and comfortable, ecological, safe and secure lifestyles
Nihon Sekkei	• Space design and optimal planning for deploying new energy devices etc.
Sumitomo Trust and Banking	• Smart town evaluation index design (environment and real estate value) • Product planning for environmentally friendly housing loans designed for Fujisawa SST
Tokyo Gas	• Installing the latest "Ene-farm" home fuel cell equipment • Proposals for comfortable and ecological living using Ene-farm
PanaHome	• Basic land readjustment project arrangements • Residential land and housing sales
Mitsui Fudosan	• Basic land readjustment project arrangements • Residential land and housing sales
Mitsui & Co.	• City block, infrastructure and real estate development also applicable for global expansion • Energy management services that take into account global smart city trends

Source Created from materials on Fujisawa SST

4.3.2 Five Areas of Value Creation in the Fujisawa SST

Energy

The first area in which Fujisawa SST is aiming for value creation is energy. Energy here means renewable energy with low environmental impact, and the energy people in the community use by adopting such energy to the maximum means values based on a concept of self-sufficiency where people produce the energy they consume as much as possible.

In Fujisawa SST, all detached homes will be equipped with solar generation systems and storage batteries. Smart home energy management systems (HEMS) will also be installed to manage electricity usage in the home. As a result, homes will not only be able to use electricity created through solar power generation to provide electricity for household electrical goods such as TVs, refrigerators, air conditioners, and personal computers but also will be able to store it in storage batteries as necessary. “Smart HEMS” cleverly make decisions about energy management based on power consumption conditions. When surplus electricity is generated, it can also sell it to an electric power company. The use of electricity in the home can be ascertained in real time via TV, smartphone, tablet, interphone or other device, thereby promoting the “visibility” of electric power.

Fujisawa SST will also provide a service for giving advice on energy to residents of detached houses based on information such as composition of family members, electricity usage, etc. By providing appropriate advice regarding excessive use of electricity or the sale of electricity, Fujisawa SST aims to promote wise use of electricity in the home. As stated earlier, through these initiatives in so-called “energy-creation, energy-saving, and energy-storage,” Fujisawa SST aims to reduce the town’s overall carbon dioxide emissions by 70 % and water consumption for everyday living by 30 % compared to 1990 levels.

After the Great East Japan Earthquake disaster in March 2011, criticism over power companies’ regional energy monopolies intensified and since then interest in energy self-sufficiency has been growing. As a project, Fujisawa SST takes into account such trends as it moves ahead with development. Efforts to promote a paradigm shift in the energy of an entire town can be considered one of its innovations. In other words, strategic innovation can be considered the capability to restructure existing industry models to create new value (Hamel 1998), and the initiative to challenge an industry model of regional monopolies by electric power companies can certainly be considered to correspond to this.

Security

Smart cities are generally recognized as environmentally-friendly cities that use electricity wisely, but cities at present also have various other problems and resolving these problems is also an important objective in the construction of smart cities. For example, since the rise of industrial cities in the 19th century up until the present, cities have achieved development by absorbing people, goods, money and information (Townsend 2013). Consequently, while many cities have grown exponentially and have prospered on the one hand, they also have many problems with their living environment. One of these is the deterioration in public safety. Major cities of the world such as New York and London are troubled by escalating crime. The creation of an urban environment where residents can live safely and securely will significantly enhance the attraction of cities.

As one of its objectives, Fujisawa SST is committed to development where residents can live safely and securely, and is reinforcing its initiatives in security. Gated towns have been established as a form of security in cities throughout the world today. In these towns, gates established at entrances and exits strictly control cars and pedestrians, thereby enhancing crime prevention. On the other hand, such an arrangement can create a feeling of being locked up and this sometimes has the adverse effect of causing unnecessary psychological pressure on residents. Therefore, instead of enclosing the town physically with gates, the Fujisawa SST project has achieved a "virtual gated town" that introduces the latest technology. In this virtual gated town, approximately 50 security cameras and lighting are effectively positioned mainly at entrances and exits to the town, public buildings, shaded areas of public parks, and major road intersections. Using LED street lights with sensors, the lights dim at night when nobody is in the area and when people or a car passes, they sense the movement, and brighten in the immediate area up to two to three steps ahead of the person or moving object. The security cameras and LED streetlights with sensors have been designed as a system to operate simultaneously.

Fujisawa SST not only relies on the latest technology of this nature but also places importance on using the "human eye" to check conditions. In other words, it attempts to achieve more reliable security through regular patrolling by guards called "security concierges."

Mobility

Mobility is one of the problem areas of cities in regard to transport. In cities in developing countries where motorization is taking place, measures to deal with the increasing number of motor vehicles are inadequate, resulting in chronic traffic congestion. At present, cars fueled by gasoline, which account for the majority of cars on the road, are also having a negative impact on atmospheric pollution and global warming.

Fujisawa SST will implement new services called "total mobility services" to address such problems. These services include a vehicle sharing service that offers electric cars, electric bike, and electric assist-bicycles as well as a rent-a-car delivery service, and a battery station service that leases charged batteries. The sharing service is a service where a user selects an electric car, electric bike or electric-assist bicycle depending on the user's circumstances and needs, how long the vehicle is to be used, or the distance to the destination, or if the user intends to rent a car, it means a rental delivery service for delivering the car to the user's home. Furthermore, to realize this as a one-stop service, a service known as the "Mobility Concierge" has been established.

For example, the Mobility Concierge can provide advice to users. Taking into consideration the distance, hours of use, and changes in the volume of traffic depending on the time of day, the Mobility Concierge can advise whether car sharing or a rent-a-car would be better, or whether an electric bicycle would be better than a car. In addition, users can check the status of car sharing or a rent-a-car booking from a television in their own home or from their smartphone. Not only that, they can also obtain various data regarding their personal use of vehicles including environmental data such as reduction of carbon dioxide, etc.

In addition, a person who uses an electric bike or electric-assisted bicycle can freely exchange and use batteries at battery stations established through the town. This relieves users from having to bother with recharging batteries after returning home, or from worrying about remaining battery power while they are commuting or shopping.

This initiative of Fujisawa SST can be said to be an initiative where companies as the service providers and residents as the customers unite in co-creation of value. Fujisawa SST has created this Mobility Service for both people who use vehicles and those who do not. Needless to say, residents include people of diverse backgrounds from children to the elderly, and therefore services that meet the needs of diverse residents must be provided. Affirmations that the services companies provide are a result of co-creation between the company and the customer rather than something unilaterally provided by the company (Prahalad and Ramaswamy 2003, 2004; Vargo and Lusch 2004) are becoming more mainstream, and the Fujisawa SST initiative can be said to be putting this principle into practice.

Healthcare

Society in Japan at present is aging at a pace unprecedented anywhere else in the world. The ratio of elderly among residents living in cities is increasing with certainty, and the problem of elderly who live by themselves and die solitary is becoming a serious social problem. Adopting "connections" as a key word, Fujisawa SST is making efforts to address this problem of social isolation. In this context, "connections" means the seamless connection of various services to people. In other words, the residents who live in Fujisawa SST, from children to the elderly, will naturally connect with one another.

In one part of the town in the Fujisawa SST is an area called Wellness Square which has a combination of elderly care facilities including a special nursing home for the elderly and serviced residential units for the elderly as well as various types of clinics, nursery and day care centers, and preparatory schools. These services are not provided individually but are seamlessly linked across conventional service boundaries to make it possible to provide optimal services to each and every resident. This is an initiative to unify medical and nursing care services so that patients can access appropriate nursing care services after they are discharged from the hospital. This is made possible through the establishment of a system known as a "comprehensive community care system." In this system, residents' health information and treatment information are consolidated and managed on a shared server through the use of ICT and can be accessed as required to provide appropriate care.

Wellness Square also has various spaces that people are free to use and where they can meet and interact with other people. By creating areas such as book corners, a science classroom for parents and children, and a consultation center that anybody from children to the elderly can use, these amenities are attempting to foster interpersonal relationships, which have a tendency to be weak among people living in cities. Interest in smart cities is often focused on "hard" areas such as establishing urban infrastructure using ICT (Rassia and Pardalos 2014), or on the viewpoint of urban planning (David et al. 2013) or innovation and competitiveness (Campbell 2012). Value creation in an area such as healthcare, however, makes people aware of the importance not only of hard aspects but also soft aspects in realizing the happiness of residents living in an urban environment.

Community

The creation of a community where people and people, and people and a town can connect with each other is the fifth area in which Fujisawa SST is aiming for value creation. Therefore, Fujisawa SST is creating services where necessary information can be accessed via a one-stop portal site. For example, it will be possible to easily access a wealth of information such as information that makes energy use "visible" in individual homes including the provision of energy-saving advice, information about events in the surrounding area as well as tourist information, information on mobility sharing reservations, word-of-mouth resident information, and confirmation of the latest conditions or safety during an emergency. This information can be accessed not only from a smartphone or personal computer but also from a smart TV, which will be a standard fitting in all detached houses. In this way, all residents from children to the elderly can obtain information that they want in every area (Fig. 4.1).

4.3.3 Process of Value Co-creation

As stated earlier, Fujisawa SST is a project led by Panasonic and strongly reflects the intentions of Panasonic from the articulation of the vision to the drafting and implementation of plans. This is because this project is the platform for putting into practice Panasonic's new management strategy to provide "comprehensive solutions for the entire house, entire building, and entire town." However, it cannot be said that Fujisawa SST is solely the project of Panasonic, and that we do not need to take into consideration the presence of other partner companies. A close look at the

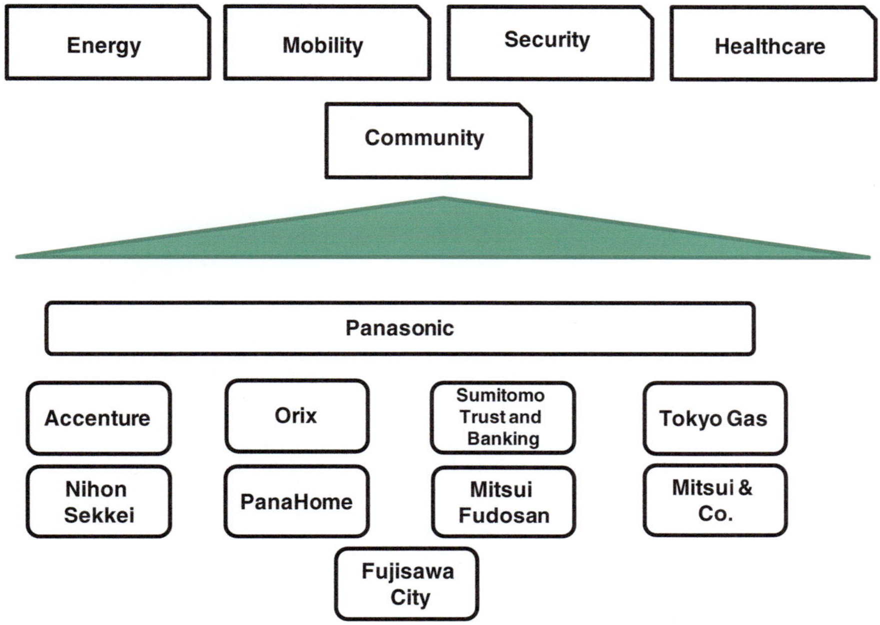

Fig. 4.1 Value creation in Fujisawa SST. *Source* Created from materials on Fujisawa SST

processes of Fujisawa SST from the formation of the project to its construction makes it clear that the intention of this project was to create new value through co-creation between Panasonic and its partner companies and Panasonic and the local government. In other words "ba" were systematically established for the Fujisawa SST project for the sake of co-creation in all processes from the conception of the vision to the drafting and implementation of the project plan to create a town. Let us now take a look at the processes of this initiative in the formation and construction period.

The land on which Fujisawa SST is located is a former Panasonic plant site. Following construction of the plant in 1961, it produced electrical goods such as black and white TVs, refrigerators and fans. In 2007 the plant was shut down and use of the vacant land had become an issue. Panasonic and the local government of Fujisawa engaged in deliberations about use of the land, and in 2010 jointly formulated the "Fujisawa SST Town Development Plan" and made it public. This plan forms the basic structure of Fujisawa SST.

The Fujisawa SST Council was then formed based on this plan to develop a town. The members of this council, which include Panasonic, its partner companies, and Fujisawa City, engaged in deliberations concerning the concept, overall goals and guidelines for developing the new town. In the course of these discussions, the Fujisawa SST concept to create a "town that brings energy to life" was born. This concept has extremely important significance in terms of the actual expression of value the Fujisawa SST intends to create. From elements based on the Fujisawa SST basic ideals of "coexistence of the environment and comfort" and "safety and security," the Fujisawa SST Council identified key elements that describe the living environment such as connections, coming together, working, learning, nurturing, health, eating, and playing, and created two slogans for the project: "A town that creates energy essential for life" and "a town that brings human vitality and energy to life." Combining these two ideas, the concept of a "town that brings energy to life" (energy meaning both electricity and vitality) came into being.

In addition, as three overall goals, the council also established environmental goals, energy goals, and safety and security goals, and set specific numerical targets for these: to reduce carbon dioxide emissions by 70 % (compared to 1990 levels), and water consumption by 30 % (compared to consumption by general facilities in 2006), source at least 30 % of all energy from renewable energy, and secure a lifeline for three days. The council also established three sets of guidelines for achieving its overall goals: the Project Design Guidelines as the guidelines governing the processes for promoting the project, the Town Design Guideline as the guidelines for designing and developing the town, and the Community Design Guidelines as the guidelines for the ongoing management of the town. All of these have been drafted by the SST Fujisawa Council after deliberation.

The final process in the initiative for forming and building the town was the establishment of a self-governing organization for the residents and a town management company. Fujisawa SST aims to become a sustainable town that will exist for more than 100 years. To achieve this, it was believed that an organization the residents could manage on their own and the presence of a company that would reflect the needs of the residents in the town services and systems were necessary. The resident self-governing organization is called the Fujisawa SST Committee. In addition to its role as a conventional self-governing committee, it has the important role of maintaining various activities relating to the environment, energy, and safety and security as well as maintaining and managing assets that it owns. This Fujisawa SST Committee will become an essential part of the residents' overall participation in the development of the town, and will generate specific ideas and activities. The Fujisawa SST Management Company, on the other hand, has the role of taking up the views of the residents expressed in the Fujisawa SST Committee, and incorporating these into the individual services and systems. The Fujisawa SST Management Company was established in March 2013 with the investment of eight partner companies, Panasonic being the largest shareholder.

As Fig. 4.2 illustrates, the initiative of the formation and construction of Fujisawa SST consisted of three phases: deliberation and formulation of the Fujisawa SST Town Development Plan (phase 1), the formation of the Fujisawa SST Council as well as deliberation and formulation of the Fujisawa SST concept, the overall goals, and the guidelines (phase 2), and the establishment of the resident self-governing organization and the town management company as well as promotion of the town's management (phase 3). After these three phases, Fujisawa SST intends to generate value in various areas through the creation of innovation. It can be said that the formation of "ba" for the creation of value through co-creation among different agents was a common feature in each of these phases.

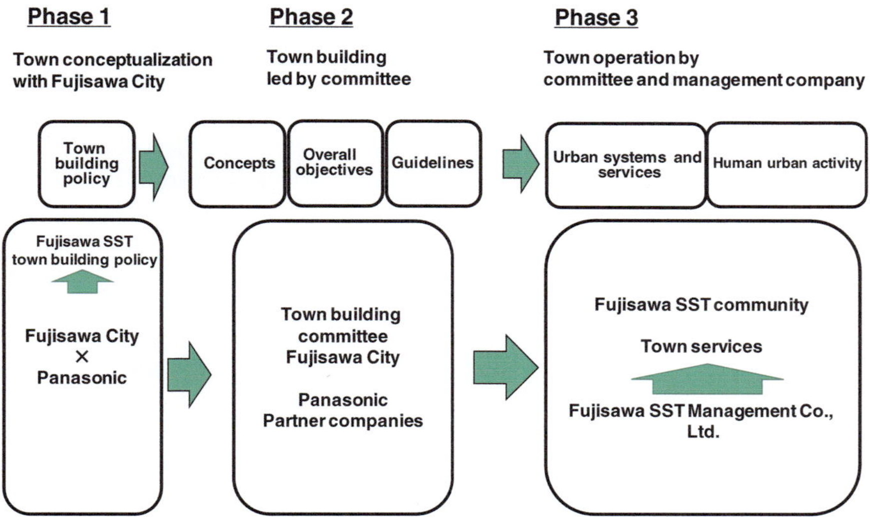

Fig. 4.2 Approaches during planning and construction of Fujisawa SST. *Source* Created by the authors from Panasonic resources

"Ba" is an important concept when discuss processes in which different agents interact and whereby value creation and innovation can be created. The concept of "ba" was first proposed in the research of Nishida (1965), a Japanese philosopher, and later developed by Shimizu (2000, 2003). Ikujiro Nonaka, in the knowledge creation theory which he himself proposed, views "ba" as a vital element in generating knowledge creation and he adopted the concept of "ba" in the area of management in earnest. (Nonaka and Takeuchi 1995; Nonaka and Konno 1998; Nonaka et al. 2008; Nonaka and Katsumi 2004; Nonaka and Tokuoka 2012). Furthermore, there is also the research of (Itami 1999; Itami et al. 2000, Itami and Karube 2004), which theorizes the formation and management of "ba." The creation of various "ba" where Panasonic, partner companies, Fujisawa City and the residents participate was evident in the processes of the formation and building of the town in the Fujisawa SST project. The agents participating in the "ba" each possess different knowledge, values, and interests and at the "ba," as these collide and merge with each other, new ideas and knowledge are created.

For example, in Fujisawa SST, every detached home is equipped with a solar generation system and storage batteries, and these are designed to link with a household fuel system known as "Ene-Farm." These systems are designed to enable the generation and storage of energy in the home. Moreover, the installation of smart home energy systems (HEMS) enables resident to manage electricity in the home wisely. Wise management means controlling the use of electricity according to needs and selling surplus power. Therefore, smart HEMS make energy "visible" to residents through a monitor that displays in real time details of electricity consumption. This information can be viewed via television at home or other device such as smartphone, tablet or interphone. There are also plans to establish autonomous symbiotic energy management where energy is efficiently used and wisely managed in the town as a whole through the coexistence of detached houses that will realize self-production and self-consumption of energy and the linking of these with building energy management systems (BEMS) installed in every facility of the town, and further developing these to community energy management systems (CEMS).

Furthermore, in the design of the city districts, boulevard trees and garden spaces will be placed along roads where breezes pass to allow the sea breezes to pass through the town. By establishing a town design guideline providing for distance of at least 1.6 meters between dwelling units, city districts were sunlight is unobstructed have been established. While such initiatives are based largely on Panasonic's state-of-the-art technology, they are achieved in collaboration with the proprietary technology and know-how of various partner companies. For example, the technology and know-how of Panasonic may be adequate for the construction of HEMS in detached houses and BEMS in commercial facilities. However, for broader area concepts involving the design of various town districts and the construction of CEMS for the overall management of the town's energy, collaboration with companies such as Accenture and Nihon Sekkei, which have knowledge in the areas of urban design and space design, is essential. When the knowledge of Panasonic collides, resonates, and co-creates with the knowledge of its partner companies, value that cannot be achieved by any single company on its own is created. What makes this possible is the "ba" of the Fujisawa SST Council.

Moreover, in the preparation of the basic plan that forms the cornerstone of Fujisawa SST, Panasonic collaborated with Fujisawa City, the local government. In short, as stated earlier, when the decision was made to build a smart town on the site where Panasonic's plant had been, a "ba" was established to discuss between them the kind of town development Panasonic and Fujisawa City wanted to see. The expectations of both parties at that time could be summed up as follows:

> Panasonic: At Fujisawa SST, we intend to put into practice our new management strategy to provide "comprehensive solutions for the entire house, entire building, and entire town" and, by gaining various kinds of knowledge, to play a role in developing the world market in the future.
>
> Fujisawa City: Through Fujisawa SST, we hope to enhance the potential of our city through the development of the naturally rich Shonan, residents with a strong awareness of the environment, and commercial and educational functions. At the same time, we hope to resolve the problems of Fujisawa City by reinforcing community disaster readiness and alleviating chronic traffic congestion.

Panasonic's discussions with Fujisawa City were ongoing since 2007, and in 2010, the Fujisawa SST Town Plan was announced. Therefore, the basic principles of Fujisawa SST incorporated into the Fujisawa SST Development Plan, such as making "a contribution to the environment and realizing a pleasant lifestyle," realizing "a safe and secure lifestyle," or becoming "a sustainable town with a smart, eco lifestyle that incorporates nature's blessings," were created through the collaboration of Panasonic and Fujisawa City.

Fujisawa SST opened to the public in 2014 and commenced in part the subdivision for detached dwellings. The final stages of development are expected to be completed in 2018 but the project at this time launched a new initiative along with the subdivision of detached houses. This was a bid to welcome new members and to create new value. As stated earlier, Fujisawa SST is a project launched by Panasonic and eight partner companies. In the lead up to the formal opening of the town in 2014, new partner companies such as SO-TWO, Gakken Cocofump Holdings Co., Ltd., Alsok Co., Ltd., and The Social Welfare Corporation Nagaoka joined the project. Adding the knowledge of new partner companies to the previously mentioned five areas of value creation Fujisawa SST is aiming for - energy, security, mobility, healthcare and community—is to create further value.

For example, SO-TWO is in charge of value creation in a commercial facility called "Shonan T-Site." Shonan T-Site creates spaces where anyone can leisurely spend time enjoying books, magazines, a café and other amenities, and is promoting the creation of a center for communicating to the world the new lifestyle developed at Fujisawa SST. SO-TWO, which is in charge of creating this new center, has experience in the construction of the Daikanyama T-SITE in Daikanyama, which is known as a center for communicating Tokyo culture. Rather than adopting the approach of conventional major commercial facilities by modifying its products in line with changes in customer needs, this company considers with shops in facilities about new lifestyles before customers' needs change, and takes the approach of continuously proposing lifestyles that are always ahead of the era. According to the company, "proposal power" is the key to the development of the Shonan T-SITE as a cultural communications center for spreading culture to the world. Therefore, rather than randomly displaying various products at the Shonan T-SITE, SO-TWO selects only those products that suit the lifestyles it proposes.

This approach of SO-TWO has the potential to bring about the creation of new value at Fujisawa SST and further enhance the town's attraction. As already indicated a number of times, the basic concept and plan for Fujisawa SST were decided with the idea that Panasonic would be at the helm, and the partner companies would be required to engage in activities that conformed to the framework of these. Therefore, SO-TWO is not permitted to engage in activities that would be in conflict with the Fujisawa SST basic plan or concept of "a town bringing energy to life," or to engage in activities that would compromise the town's image. Nevertheless, although there are restrictions in the overall framework, in individual areas mechanisms have been established that allow for partner companies to utilize their knowledge as much as they wish. It can be said that initiatives at creating new value at the Shonan T-SITE are effectively left to SO-TWO. In this way, at Fujisawa SST there is a mixture of top-down methods by Panasonic and bottom-up methods where partner companies utilize their own knowledge, and a mechanism is in place whereby new knowledge is created from the merging of knowledge of the two.

4.4 Construction of a New Business Model

Companies participating in the Fujisawa SST project belong to diverse industries including electronics, housing, gas, real estate and finance, and not one of these companies is a so-called specialist in urban development. In fact, there is not even one company whose core business is urban development. However, the project is an opportunity for each of these companies to develop new business. In particular, Panasonic, which is leading the project, can be said to be attempting to create a new business model based on knowledge it gains from the Fujisawa SST. The next sections will take a look at Panasonic and PanaHome and examine their structuring of a new business model.

4.4.1 Panasonic

It goes without saying that Panasonic, which is at the helm of the Fujisawa SST project, is a major electronics manufacturer representative of Japan. The mainstays of its business are home electronics products such as televisions, video recorders, refrigerators, washing machines and air conditioners, and in recent years the company has been involved in the production and sales of solar panels and automotive electronics. As the commodification of its mainstay products progresses, however, cutthroat price competition among rival companies in South Korea and China has escalated. It cannot be denied that Panasonic is at a competitive disadvantage in terms of costs, and the deterioration its profits have suffered in recent years is no secret.[2]

It is against this backdrop that the Fujisawa SST project was planned. For a company that had been involved in the production and sales of home electronics, a project like Fujisawa SST, which was to develop an entire town from scratch, was a wholly new experience, and it can be said that Panasonic made a very bold move to branch out into this new business. Leading up to this development was the increasingly fierce price competition in the area of home electronics. Under the present circumstances where both South Korean and Chinese companies have grown in strength, Panasonic, would have been hard put to improve its earnings on the basis of individual products. Moreover, for a company like Panasonic, which was involved in a wide range of products and businesses, town development business like the Fujisawa SST project seemed appropriate as a new business for mobilizing the knowledge assets it possesses. The existence of social issues, which require solutions such as making the transition to a low-carbon society and addressing various problems that cities contend with also offer new business seeds for the company.

[2]This situation is not limited to Panasonic alone but also applies to other electronics manufacturers such as Sharp and Sony.

In the history of Panasonic, which will celebrate its 100th anniversary in 2018, Fujisawa SST can be considered a groundbreaking turning point in business. In short, Panasonic is attempting to change its existing business model as a manufacturer of individual white goods and devices to a creator of new urban spaces in the 21st century. "Solutions for the entire home, entire facilities, and entire towns," the new strategy concept that Panasonic adopted certainly symbolizes this. Furthermore, a characteristic of this new business model is the value to be created from business is not based on the efforts of a single company but on the efforts of a number of companies that generate value through co-creation unachievable by a single company. In other words, instead of competing over the performance of individual products, the individual companies, through the exchange and merging of knowledge and know-how they possess respectively, create value through co-creation.

A further point must also be noted in regard to Panasonic's new business model in Fujisawa SST. That is the viewpoint of the solutions business. Fujisawa SST is a project based on a vision of 100 years of development. In other words, once construction of the town is complete, the project does not end there. Fujisawa SST is a project that has adopted as its mission development for the next 100 years. Therefore, from completion of the town in 2018 until the year 2108, the project will engage in ongoing management of the town, with plans for management during that time organized into periods of 30 years: growth period (2018–2048), maturation period (generation change) (2048–2078), and evolution period (generation change) 2078–2108. Needless to say, various changes in the environment will occur with the passage of time. Facilities that were initially considered state-of-the-art will most likely be obsolete after 30 years or 50 years. Generational changes will also take place with residents who live in the city with the passage of time and, in tandem with these changes, it can be assumed that needs in the living environment will also change. Failure to respond appropriately to such changes could result in the gradual deterioration of the town, and even lead to its demise after 100 years. Therefore, to realize a town the will continue for the next century requires responding to the passage of time and providing solutions for the various issues that arise. In other words, the so-called solutions business will play a vital role. The solutions business itself is not particularly new and many companies already engage in this business, but there are few cases where companies have been challenged to provide long-term solutions of 100 years. For a home electronics manufacturer like Panasonic, whose main business model to date has been to develop new products in the short term to generate profits, this is indeed a major shift.

Hamel and Prahalad (1989) argue that an underperforming company must reconsider its strategy concept and construct a new strategy model in order to regain its former freshness and vigor. It can be said that this is exactly what Panasonic has done. Nevertheless, changing a business model is not something that is easy to do. This is because a change in business model is generally accompanied by considerable pain including downsizing and consolidation of existing businesses and attendant redundancies in personnel. In the early 2000s Panasonic also attempted to change its business structure by adopting a policy of "scrap and build" but the company continued to suffer from poor earnings. It can be said that Fujisawa SST represents an innovation in strategy (Markides 1998) that will dramatically change the existing business of Panasonic.

4.4.2 PanaHome

PanaHome Corporation is a group company of Panasonic and housing-related business is its core business. The company's main business areas are the construction of detached houses and rental complex housing, renovation construction contracting, sale of land and buildings for subdivision and condominiums, real estate agency business, and lease management. As a company of the Panasonic group, PanaHome plays a central role together with Panasonic in the Fujisawa SST project. However, its portfolio in the project goes beyond the conventional role of a housing developer and includes new business areas.

As Table 4.1 shows, the role required of PanaHome in Fujisawa SST is infrastructure arrangements for development including land reallocation, participation in housing land and residential subdivision business, urban planning, urban design, establishment of rules for development, and the creation of service business schemes for the maintenance and management of the town. A role with the very same portfolio has also been assigned to Mitsui Fudosan. In other words, both companies are expected to cooperate in fulfilling a role with the same portfolio. This role includes participation in housing land and residential subdivision business, an area of business PanaHome has been involved in for some time but urban planning, urban design, establishment of rules for development, and the creation of service business schemes for town maintenance and management are new areas that PanaHome had not engaged in previously. Panasonic's intention in assigning such a role means that in addition to fulfilling the role as a housing developer, PanaHome is being asked to assume the new role of planner of urban development and the role of a management company responsible for the management of the town. This will provide new business potential for PanaHome as a housing developer and will lead to further enhancement of its added value. In other words, in the Fujisawa SST, PanaHome will not only perform a role as a developer of detached housing and residential subdivision, its business areas to date, but will also become actively involved in new areas such as urban design, maintenance and management where it will be able to amass valuable experience. As a result, there is also a possibility that PanaHome will make a transition from its existing business model as a simple housing developer to a new business model as an urban development planner.

The ultimate goal of the Fujisawa SST project, as stated earlier, is to develop markets in Asia based on the knowledge gained from this project. The aim is to establish a so-called "Fujisawa Model" to build smart cities in regions of Asia. At that time, PanaHome intends to engage in market development not as a housing developer but as an urban development planner with a new business model.

4.5 Implications

To conclude this chapter, I would like to summarize implications drawn from this case analysis. First of all, I would like to consider implications that can be drawn from an analysis of the "ba" where Panasonic along with its partner companies and Fujisawa City discussed the formation and building of Fujisawa SST. As already indicated, Fujisawa SST was a project initiated by Panasonic, and every process from project planning to implementation of the project proceeded according to the strategy envisioned by Panasonic. Under such circumstances, even if a "ba" was formed to discuss urban development, it would appear that the roles of the partner companies and Fujisawa City would be simply to rubber stamp the plans indicated by Panasonic, and that their participation in "ba" would simply be a formality. Confirmation of "ba" formed in the case of the Fujisawa SST, however, shows that this was not the case and that the "ba" effectively fulfilled the function of deliberation. For example, to formulate the urban development plan, Panasonic engaged in deliberations with Fujisawa City for a period of three years before they jointly announced a plan. If the plan were to be unilaterally formulated on the basis of Panasonic's strategy, it should not have required spending such a lengthy time in deliberation. Therefore, it can be assumed that in its planning Panasonic was well aware that the cooperation of the local government of Fujisawa City would be essential for the success of the Fujisawa SST project.

Furthermore, to draft a town concept, the overall goals, and guidelines, the Fujisawa SST Council was established as a "ba," which in addition to Panasonic included all the partner companies and Fujisawa City as members. It was in this "ba" that important matters relating to value creation in Fujiwara SST were discussed. According to Panasonic, the drafts of the concept, overall goals, and guidelines were presented by Panasonic and in the course of deliberation with the partner companies and Fujisawa City in the council, they collectively worked out a final proposal. What confirmation of these results indicates is that "ba" formed in the Fujisawa SST project were not in any way just formal "ba" for rubber stamping the intentions of the leader company. They were "ba" that demonstrated flexibility. While respecting the intentions of the leader company, these "ba" also adopted the opinions of other participating parties with a view to creating value of a higher dimension. The fact that "ba" of this nature were formed in the initial stage of the town's formation and construction is evidence that Panasonic believed nurturing a relationship of trust in the early stages (Kanter 1994) was essential to succeed in collaborations with the partner companies.

What, then, can be said about management of the "ba?" Various "ba" were established for deliberating on town development of Fujisawa SST but it was the leader company Panasonic which solely performed the role as leader. As stated before, deliberation concerning the concept, overall goals, and guidelines of Fujisawa SST took place in the Fujiwara SST Council. However, this process began with Panasonic's presentation of drafts for these and, based on these, final drafts were completed after repeated deliberations with the partner companies and

Fujisawa City. Since Panasonic existed as the definitive leader in the Fujisawa SST project, it was clear which agent was to lead the "ba." Thus, there was no concern over lack of clarity as to the party that would take leadership of the "ba," which can be problematic when relationships among members are complex. Furthermore, Panasonic's management of the "ba" can be said to be a balanced mix of top-down methods initiated by Panasonic and bottom-up methods where knowledge of the partner companies was taken on board. This management can also be described as management that leads to value creation. For example, the concept of Fujisawa SST, "the town that brings energy to life," is the product of the co-creation of Panasonic, the partner companies and Fujisawa City, and communicates to the rest of the world the value of the town as a whole in the form of a clear message. On the other hand, the five values to be individually created (energy, security, mobility, healthcare, and community) by Fujisawa SST were established by Panasonic based on its company strategy. In this way, the value to be created by Fujisawa SST is clearly indicated, and this can be said to be the result of sound management of the "ba" on the part of the leader company. In recent years, studies have been conducted regarding the kind of management that is required when people of various backgrounds and firms of different industries collaborate (Kodama 2009, 2010), and the method at Fujisawa SST of differentiating the use of top-down and bottom-up method according to circumstances can be considered one viable approach.

The next question to consider is what implications there are in regard to "emergence" at Fujisawa SST. The existence of a self-motivated "ba" and a bottom-up process are the assumed conditions for the manifestation of "emergence" but the environment of Fujisawa SST did not necessarily conform to these conditions. Nevertheless, emergence manifested at the individual workplace level can perhaps be clearly seen in areas such as the activities of SO-TWO at the Shonan T-SITE mentioned earlier. While the overall plan and framework were decided in a top-down manner by the leader company, and the partners follow these accordingly, in individual areas the leader company refrains from getting involved as much as possible to enable the partner companies to demonstrate their knowledge to the fullest. In fact, it can be said that emergences resulting from the merging of top-down and bottom-up initiatives is a characteristic feature of Fujisawa SST. Therefore, for the leader company, emergences achieved in this way are intended emergences that can be predicted to some extent. This is because they are manifested within the plans and the strategy framework the leader company itself decided on. In addition to this, there are various mechanisms in place in the Fujisawa SST project for exploiting bottom-up emergence such as the Fujisawa SST Community, the residents' self-government organization, and the town management company, which takes on board the various opinions and requests of residents brought up in their committee meetings and realizing them in some form. Therefore, it can be assumed that various forms of emergence will occur as the town develops.

What will connect these diverse emergences into value creation for the town as a whole will be "synthesis." Nevertheless, it cannot be said that validation of this aspect is adequate. At present Fujisawa SST is a project currently in progress and, unfortunately, it is not possible to confirm "synthesis" as an element that is evident in the project at present. A hypothesis such as that described in the "analytical framework," that is, a scenario where the process of "intended value creation" based on the intended strategy of the leader company and the process of "unintended value creation" that springs up at the workplace level merge and create new value is quite fascinating but at present is a phenomenon that has not been ascertained. Nevertheless, one assertion that can be made regarding the Fujisawa SST project is that synthesis will occur without fail. This is because it is clear who the agent responsible for synthesis is. In this case, of course, that is Panasonic. For example, even if various emergences occur in individual areas, and among these are elements that would lead to overall value creation, unless there is an agent that fully comprehends these behaviors and links them to value creation, overall value will not be created. Under ordinary circumstances, the leader of the "ba" fulfills this role and if a clear leader of the "ba" exists and that leader has adequate capability, synthesis will occur without fail. I intend to continue to observe this aspect in the future.

References

Andrews, K. R. (1971). *The concept of corporate strategy*. Homewood, III: Dow Joes-Irwin.

Ansoff, H. I. (1965). *Corporate strategy: An analytical approach to business policy for growth and expansion*. New York: McGraw-Hill.

Campbell, T. (2012). *Beyond smart cities: How cities network, learn and innovate*. New York: Routledge.

David, G., Rob, K., & Gordon, M. (2013). Grappling with smart city politics in an era of market triumphalism. *Urban Studies, 50*(11), 2151–2157.

Hamel, G. (1998). Strategy innovation and quest for value. *Sloan Management Review, 39*(2), 7–14.

Hamel, G., & Prahalad, C. K. (1989). Strategic intent. *Harvard Business Review,* 63–76.

Itami, H. (1999). *Management of 'Ba': A new management paradigm*. Tokyo: NTT Publishing.

Itami, H., & Karube, M. (Eds.). (2004). *Strategy and logic of invisible assets*. Tokyo: Nihon Keizai Shinbun-sha.

Itami, H., Nishiguchi, T., & Nonaka, I. (Eds.). (2000). *The dynamism of 'Ba' and the company*. Tokyo: Toyo Keizai Shinpou-sha.

Kanter, R. (1994). Collaborative advantage: Successful partnerships manage the relationships, not just the deal. *Harvard Business Review, 72*(4), 96–108.

Kodama, M. (2009). *Innovation networks in knowledge-based firms: Developing ICT-based integrative competences*. Cheltenham, UK, Northampton, MA, USA: Edward Elgar.

Kodama, M. (2010). *Boundary management: Developing business architectures for innovation*. Heidelberg, Dordrecht, London, New York: Springer.

Markides, C. (1998). Strategic innovation in established companies. *Sloan Management Review, 39*(3), 31–42.

Mintzberg, H. (1973). Strategy-making in three models. *California Management Review, 16*(2), 44–53.

Mintzberg, H. (1978). Patterns in strategy formulation. *Management Science, 24*(9), 934–948.

Mintzberg, H. (1990). The design school: Reconsidering the basic premises of strategic management. *Strategic Management Journal, 11*, 171–195.

Nishida, K. (1965). *A full of collection of Kitaro Nishida's work*. Tokyo: Iwanami Shoten.

Nonaka, I., & Katsumi, A. (2004). *The essence of innovation*. Tokyo: Nikkei BP-sha.

Nonaka, I., & Konno, N. (1998). The concept of 'Ba': building a foundation for knowledge Creation. *California Management Review, 40*, 40–54.
Nonaka, I., & Takeuchi, H. (1995). *The knowledge creating company*. New York: Oxford University Press.
Nonaka, I., & Tokuoka, K. (Eds.). (2012). *Business model innovation*. Tokyo: Toyo Keizai Shinpou-sha.
Nonaka, I., Toyama, R., & Hirata, T. (2008). *Managing flow: A process theory of the knowledge-based firm*. London: Palgrave macmillan.
Prahalad, C. K., & Ramaswamy, V. (2003). The new frontier of experience innovation. *Sloan Management Review, 44*(4), 12–18.
Prahalad, C. K., & Ramaswamy, V. (2004). Co-creation experience: the next practice in value creation. *Journal of Interactive Marketing, 18*(3), 5–14.
Rassia, S Th, & Pardalos, P. M. (Eds.). (2014). *Cities for smart environmental and energy future: Impacts on architecture and technology*. Heidelberg, New York, Dordrecht, London: Springer.
Shimizu, H. (2000). *'Ba' and co-creation*. Tokyo: NTT Publishing.
Shimizu, H. (2003). *The concept of 'Ba'*. Tokyo: University of Tokyo Press.
Steiner, G. A. (1969). *Top management planning*. New York: Macmillan.
Townsend, A. M. (2013). *Smart cities: Big data, civic hackers, and the quest for a new utopia*. New York, London: W.W. Norton and Co.
Vargo, S. L., & Lusch, R. F. (2004). Evolving to a new dominant logic for marketing. *Journal of Marketing, 68*, 1–17.

Chapter 5
Co-creation of Value Through Collaboration of Government and Companies—Case Study of the Yokohama Smart City Project

5.1 Introduction

In this chapter, I will take up the case of the Yokohama Smart City Project (YSCP), a smart city demonstration project conducted in the city of Yokohama. Implemented as one of the projects for the Next-Generation Energy and Social Systems Demonstration project sponsored by the Ministry of Economy, Trade and Industry (METI), the project was initiated entirely by the government. In addition to the city of Yokohama, similar projects were implemented in Toyota City, Keihanna Science City, and the city of Kita Kyushu. Among these, the Yokohama City Project was the largest. A city with a population of 3.7 million people, Yokohama is renowned as one of the world's most thriving metropolises. It also has a renowned international trade port, the Port of Yokohama, which is one of Japan's largest commercial and industrial hubs. The Next-Generation Energy and Social Systems Demonstration aims to transform the existing metropolis into a smart city that already has infrastructure, and its main objective is to obtain various data through demonstrations. Participating in the YSCP are major companies representative of Japan such as Toshiba, Hitachi, Panasonic, Nissan Motor, TEPCO and Orix, and it is expected that new value will be created through the integration and resonance of the technologies and know-how the respective companies have.

5.2 Analytical Framework

As in Chaps. 3 and 4, I would first like to confirm a number of points of note in the case analysis in this chapter too. The first point that must be mentioned is that the Yokohama Smart City Project (YSCP) to be examined in this chapter was a project initiated by the government. Both of the cases taken up in Chaps. 3 and 4 were projects initiated by private sector companies, and while the local governments also

© Springer Japan 2016

N. Tokoro, *The Smart City and the Co-creation of Value*,
SpringerBriefs in Business, DOI 10.1007/978-4-431-55846-0_5

participated and their involvement was acknowledged to a certain degree, their roles were limited, and neither of these projects was initiated by the government. On the other hand, the project examined in this chapter is one that was initiated solely by the government, and was executed according to a master plan prepared by the government, which made clear provisions for the project budget, project period, objectives of the various demonstrations to be conducted during the project implementation period, and the respective roles of the participating companies. Therefore, the companies participating in the project were to conduct their activities within the framework of their defined roles according to the principles of the master plan established beforehand. In other words, the participants' exercise of discretion in the project activities would be limited. In this respect, the case studied here differs significantly from the cases presented in Chaps. 3 and 4. Therefore, the factors described above can be expected to have more than a small impact on the analytical framework in this chapter.

In short, there is a possibility that the analytical framework of the Yokohama Smart City Project will not conform to the theoretical framework of a relationship-based strategy established as the framework in the case analyses of the smart cities in this book, which fundamentally consists of a framework of highly autonomous "ba," or shared context in motion, emergence originating in bottom-up environment, and synthesis that consolidates emergences and leads to value creation. This is because the establishment of a highly autonomous "ba" cannot be expected in an environment where the budget, project implementation period, and division of roles of the participating members are clearly defined beforehand and, therefore, expectations of bottom-up emergence will be modest at best. Furthermore, there is a strong possibility that value creation through synthesis will not be manifested. In other words, there is a strong possibility that no beneficial implications will be drawn even if a theoretical framework based on the assumption of a relationship-based strategy is applied to this case study. This leads to the question as to what kind of analytical framework should be applied in the case analysis in this chapter under these circumstances.

Given the nature of the project, that is, a project where the preparation of plans and the roles of the respective companies have been clearly defined by the government, the establishment of an analytical framework in line with the tenets of advocates of strategic planning (Ansoff 1965; Steiner 1969; Andrews 1971) may be considered. In short, this means the establishment of an analytical framework reliant on characteristics such as plans drafted by top management, top-down, systematic, rational execution of plans, and clarification of every person's role, among others. In such a case, any "ba" formed in the course of the project would be a "ba" where autonomy is restricted and where the creation of any value would not be through emergence or synthesis but through the expression of value intended at the planning stages. This leads to the next question, that is, in the case analysis in this chapter how value intended at the planning stage will be validated in the course of the top-down execution of the project. In this case, rather than synthesis, an analytical approach will be adopted for the analytical approach.

Co-creation is fundamentally an event manifested through synthesis. Therefore, adopting an analytical approach may be considered inappropriate for analyzing the co-creation of companies hailing from different industries. On the other hand, it is not possible to clearly separate analysis and synthesis. Put another way, it is sufficiently possible to identify synthesis within analysis. For example, in the case of the Yokohama Smart City Project, even if the project progresses systematically in a reasonable manner according to the master plan where individual companies have predetermined roles, some form of emergence may occur in the process and develop into synthesis, resulting in the creation of value not initially intended. Emergence is said to take place easily in organizations of Japanese companies (Nonaka and Takeuchi 1995). Therefore, even under top-down or restricted "ba" conditions, the possibility that emergence and synthesis will occur cannot be ruled out. Whatever the case, the main focus of the analysis in this chapter will be in determining whether co-creation of value will occur under the given circumstances and, if so, what mechanism causes this to occur.

5.3 Yokohama Smart City Project (YSCP)

5.3.1 Project Missions and Basic Principles

In January 2010, the Ministry of Economy, Trade and Industry (METI) selected four areas as demonstration areas for its Next-Generation Energy and Social Systems Demonstration project, and the Yokohama Smart City Project (YSCP) was selected as one of these. Accordingly, the YSCP subsequently conducted demonstrations in technology, systems and a business model for building a smart city in Yokohama. The project implementation period was for the five-year period from 2010 to 2014.[1]

The missions of the Yokohama Smart City Project can be roughly divided into two areas. The first, needless to say, was to establish a low-carbon society. Therefore, the first mission was to realize in the city of Yokohama, one of Japan's leading metropolises, a low-carbon society with reduced carbon dioxide emissions. This was to be achieved by using the latest technologies to transform on a large scale the city's social infrastructure in areas such as electric power and transportation, among others. The second mission was to establish a smart city model of the world's highest level based on various data and know-how obtained from the demonstrations in Yokohama, and to capture overseas demand and increase the national wealth by participating in the construction of social infrastructure overseas, particularly in emerging countries of Asia where growth is marked. The Yokohama Smart City Project is an initiative to convert a large metropolis with a population of

[1]For details regarding the Yokohama Smart City Project, the author relied on information in the Master Plan prepared by the Yokohama City Climate Change Policy Headquarters in August 2010.

3.7 million people to a smart city. However, efforts to transform existing social infrastructure where residents are actually living is no easy task. Campbell (2012) argues that ongoing study and the creation of innovation are essential when building a smart city, and this viewpoint is all the more applicable when an existing city is to be transformed into a smart city. At the same time, knowledge gained from accumulating such experience becomes a valuable asset, and enables an organization to provide beneficial solutions for building social infrastructure in emerging countries in Asia, which suffer from various urban problems. China in particular, which suffers from population migration to cities and environmental pollution, has shown strong interest in the building of smart cities, and is a promising candidate as an export destination for Yokohama-type solutions.

To achieve the above missions, the Yokohama Smart City Project has identified four elements as requirements for a smart city: a smart city should be scalable, speedy, sophisticated and satisfaction. Scalable, the first of the elements, takes into account the size of cities in emerging countries of Asia, which are assumed to be the future export destinations of Yokohama-style solutions. Asian countries such as China and India have enormous populations as well as cities of a significant size. Therefore, it is likely that data and know-how obtained from demonstrations of small smart cities may not be applicable to such cities. The demonstrations in the city of Yokohama are of the largest scale in the world, and a requirement of this project is to verify that findings from demonstrations are sufficiently relevant to other large cities in the world in terms of size.

Speedy, the second element, means the project will aim for the "speedy" development of smart cities. It can be said that the rapid construction of smart cities is essential in emerging countries, which are growing rapidly. Therefore, the Yokohama Smart City Project considers accelerating the launch of smart cities by introducing "urban package solutions" which include both construction and the operation of facilities, as well as utilizing existing urban infrastructure to the maximum extent possible.

The third element, sophisticated, means the project will aim for a smart city that achieves both excellent cost performance and high quality by not only promoting the latest technologies that Japan has to offer but also integrating these technologies with inexpensive mature technologies. Furthermore, another aim of the project will be to transform large cities to smart cities that have a spirit of innovation by encouraging people to be more forward-looking in their everyday lives. This will be achieved not only by changing urban infrastructure but also inducing changes in people's ways of living.

The last of the four, satisfaction, means promoting an orientation toward solutions where the citizens of a smart city willingly participate in efforts to achieve ecological living. This means to develop an environmentally friendly lifestyle that does not compromise convenience or impose restraints.

The above are the basic principles of the project but in addition to these, the city of Yokohama has established the Yokohama Climate Change Action Policy, which commits to the following goals.

(1) Reduce per capita greenhouse gas emissions to at least 60 % of the fiscal 2004 level by 2050.
(2) First reduce per capita greenhouse gas emissions to at least 30 % of the fiscal 2004 level by 2025 to achieve the above target, and introduce the use of renewable energy 10-fold from the fiscal 2004 level.

To achieve these ambitious targets to reduce carbon dioxide, the project has committed to the basic principle of developing social systems that utilize low-carbon technologies in the three areas of energy, buildings, transport and transportation. To achieve these goals, the city has also made a commitment to make efforts based on the following three steps.

(1) Demonstrate "technologies" by actively adopting technologies in an integrated platform one step ahead of their practical application.
(2) Demonstrate the "economic performance" of technologies through the trial introduction of service systems and new business for promoting the dissemination of validated technologies.
(3) Demonstrate the "effects of dissemination" of technologies as a social system through the broad dissemination of services that have been established (including linking these with system design).

In other words, the city of Yokohama made a commitment to proceed with the project to achieve the above goals while establishing systematic, rational demonstration processes for technologies, economic performance, and dissemination effects, and while bearing in mind the four requirements of a smart city mentioned earlier: scalable, speed, sophisticated and satisfaction. As indicated earlier, transforming an existing city with social infrastructure already in place to a smart city is no easy task. Unlike building a new city on vacant land where nothing exists, in this case, the project had to take into consideration the impact work would have on the activities of citizens who actually live in the city as well as companies which engage in economic activities. Therefore, the transformation of an existing city to a smart city, as indicated by Hollands (2008), is not simply a process that involves "hard" aspects of merely introducing smart infrastructure using ICT but also requires the establishment of "soft" aspects such as the building of creative partnerships, the transfer of knowledge and the development of capabilities, among others. Moreover, when an existing city makes the transition to a smart city, the issue of governance of the city must be taken into consideration (Deakin 2014).

The Yokohama Smart City Project demonstrations were to be implemented in three areas: the Minato Mirai 21 area, the Kohoku New Town area and the Yokohama Green Valley area. Of these, the Minato Mirai 21 area has a population of approximately 7000 and approximately 3600 households, and is a large commercial district with high-rise buildings and commercial facilities. During the demonstration, building energy management systems (BEMS) were to be introduced in each skyscraper business building in the area, and the project was to achieve management of a group of existing and new BEMS by constructing an integrated BEMS to control the BEMS in the respective buildings. Moreover, the "visualization" of energy used in the buildings and optimization of energy management in the business buildings including heating and air conditioning were to be validated. Furthermore, in the transportation system in the area, infrastructure for prompt transfer services was inadequate. Bearing this in mind, the project was to develop a low-carbon mobility infrastructure through the introduction of EV car sharing and recharging stations for business purposes as well as EV circuit buses, and to validate this transportation system with a low environmental load.

Kohoku New Town Area is a bed town of the city of Yokohama with a population of approximately 200,000 and 75,000 households. Utilizing the characteristics of the area, the project was to tap into demand for home improvements and validate low-carbon improvements based on the introduction of insulation retrofitting, home energy management systems (HEMS), and residential photovoltaic (PV) systems, etc. Other areas that were to be validated included the introduction of renewable energy at public facilities and parks, the introduction of BEMS when supply-side equipment and demand-side equipment were to be updated in district heating and cooling (DHC) in the area in front of the railway station, and the integration of the transport of high-temperature waste heat from the waste treatment plant and a DHC system.

With a population of approximately 210,000 and approximately 87,000 households, Kohoku Green Valley area is a compact area along the waterfront with housing estates, industrial estates, and public facilities including schools and hospitals. In this area, the aging of the population is progressing and the birthrate is declining. Therefore, with the cooperation of industry, government, the academic community, and local residents, the project was to promote the revitalization of the local economy as well as the introduction of smart technologies in housing and industrial estates through the introduction of renewable energy and highly efficient equipment. In addition to introducing smart technology, the project also aimed to work towards the establishment of a low-carbon model area by raising awareness of the environment among local residents and businesses to ensure that pro-active energy-saving activities would be widely practiced (Fig. 5.1).

Details of the seven demonstrations implemented during the YSCP are given below.

5.3.2 Seven Demonstrations

Large-Scale Introduction of Renewable Energy

This demonstration was to verify the hypothesis that the intensive installation of renewable energy equipment in a specific area and the establishment of an energy management system in the area using CEMS combined with storage batteries, etc.,

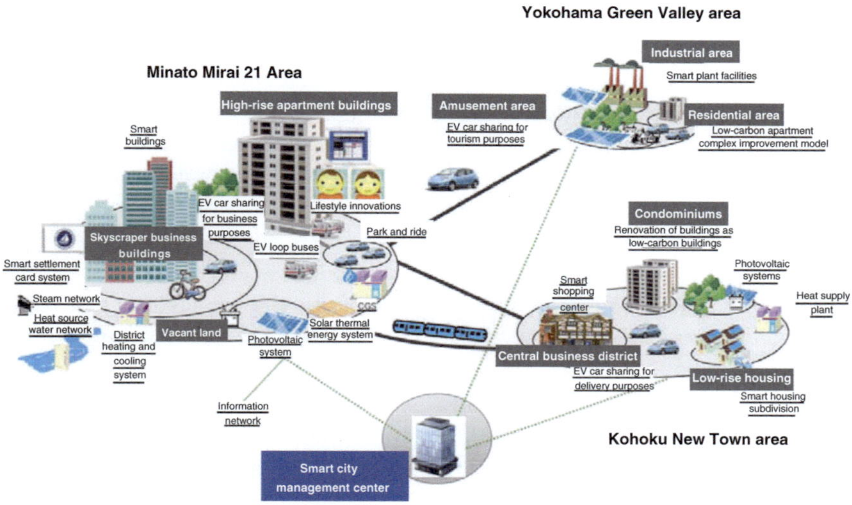

Fig. 5.1 Future vision of the Yokohama Smart City Project. *Source* Excerpt from the social system based on next-generation energy demonstration, Yokohama Smart City Project master plan

would achieve both a stable power supply and quality of life for the residents, thereby paving the way for the large-scale introduction of renewable energy. To be specific, solar power systems were to be introduced into the three areas—Minato Mirai 21 area, Kohoku New Town area, and Yokohama Green Valley area. The goal was to introduce renewable energy with a total capacity of about 27 MW including about 13 MW of residential PV systems (about 4200 households × about 3 kW) and about 14 MW of medium and large PV systems on the assumption that this would bring the percentage of power generated by residential PV systems to more than 5 % of the final energy consumption by the households in the demonstration area.

Furthermore, the demonstration was also to install solar thermal panels, solar thermal absorption water heater/cooling systems for air conditioning, and gas engine water heaters, etc. in buildings in the three areas, and to conduct demonstrations to verify technology and the effect on carbon dioxide reduction. Demonstrations were also to be conducted on the introduction of river water source heat pumps to verify their effect on carbon dioxide reduction. A river water source heat pump is a highly efficient heat source system for air-conditioning which utilizes river water with stable temperature throughout a year (i.e. the water stays cool in the summer and is warm in the winter relative to the air temperature) as heat source water or cooling water in order to provide heating or cooling for buildings. This demonstration was to verify the hypothesis that the introduction of river water source heat pumps in new buildings to be constructed in the area around Yokohama Station where abundant river water is available would contribute to a significant reduction in CO_2 emissions through the replacement of fossil fuel (used in boilers).

Home Energy Management Systems (HEMS)

This demonstration was to verify the hypothesis that both the quality of life (QOL) of residents and a reduction in household CO_2 emissions could be achieved through the demonstration (at a certain level) of the effects of load creation and load shifting through the streamlining and visualization of household power consumption using HEMS and the introduction of economic incentives which would contribute to the introduction of a large number of photovoltaic systems, and the utilization of highly efficient energy systems effective in the reduction of CO_2 emissions.

Appropriate methods to approach potential users who are highly environmentally aware were also to be considered as a means of disseminating HEMS. Participation in the demonstrations was to be by public invitation. As an incentive for the introduction of HEMS, installation of the system was to be combined by the introduction of equipment that could be effective on its own such as PV systems and solar water heaters. An additional subsidy was also to be provided by the city of Yokohama for the installation of PV systems if the user agreed to participate in the project and connect his/her system to CEMS. In regard to new detached houses, houses already equipped with HEMS were to be offered through a collaboration between home builders and HEMS manufacturers.

In condominiums, solar thermal energy systems, PV systems, fuel cells and storage batteries were to be installed, and demonstrations were to be conducted on the optimal control of multiple decentralized power supply systems among houses and energy-consuming equipment.

Building Energy Management Systems (BEMS)

This demonstration was to verify the hypothesis that the use of BEMS would be promoted and CO_2 emissions from business and commercial buildings would be reduced through the provision of various combinations of BEMS and highly efficient energy systems, the rating of buildings in accordance with their environmental performance, and the combining of systems with services that would reduce building owners' initial costs and maintenance costs.

The functions of "coordination with CEMS and integrated BEMS" and "visualization of CO_2 emissions" were to be added to BEMS as standard functions for newly installed BEMS and as additional functions for existing conventional BEMS in order to verify the effect on CO_2 emissions reduction at each building, in each group of buildings and in each area. At the time of the BEMS installation, some units were to be installed with highly efficient energy systems and storage battery systems. To achieve further CO_2 emissions reduction, optimum operation of these systems was then to be achieved using BEMS connected to CEMS. Furthermore, service models and measures were to be considered to promote the dissemination of BEMS. The project was also to consider schemes for reducing users' initial investment costs for the introduction of BEMS as well as the creation of a mechanism that would enable comparative assessments and rating of the environmental performance of buildings through visualization based on final utility. Moreover, the demonstration was to consider solutions that would contribute to the reduction in building management and operating costs.

Thermal Energy Management at the District Level

Four waste treatment plants are currently in operation in the city of Yokohama. Each plant recovers thermal energy from waste incineration by converting it into steam, which is then used in steam turbine generators or for heat supply to nearby facilities. Although waste heat was already being supplied to nearby facilities, it was not being supplied to other areas. Therefore, the project was to conduct feasibility studies to examine the environmental performance, technical feasibility, and project profitability of the construction of supply pipelines from the respective waste treatment plants to neighboring heat demand areas, and to verify the hypothesis that CO_2 emissions could be cut significantly by utilizing this high-temperature underutilized energy.

Furthermore, in the Minato Mirai 21 area, heat generation units were to be installed, a district energy management system was to be established, and the effective system together with the expansion and increase in plant facilities was to be verified. In the Kohoku New Town area, BEMS was to be installed at the district heating and cooling (DHC) facilities (the supply side), and at the facilities receiving the district heating and cooling services (the demand side), the technical verification of the effect on CO_2 emissions reduction was then to be conducted.

Mutual Supplementation Between Community Energy Management Systems (CEMS) and Large-Scale Power System Networks

This was to be a demonstration for connecting CEMS to HEMS, which are to be installed in detached houses and condominiums in the area, and BEMS, which are to be installed in business buildings, and car sharing facilities that utilize fast charging stations and chargeable/dischargeable EVs. CEMS will send commands to HEMS, BEMS and car sharing facilities and collect energy management data. In areas where PV systems have been introduced intensively, operational planning of PV output and storage batteries were to be established, and in cases where large numbers of PV systems have been installed in the area, CEMS were to indirectly control supply and demand through HEMS and BEMS to accommodate local demand on the day with PV output.

Next-Generation Transport Systems

In addition to developing charging infrastructure in preparation for the introduction of large numbers of EVs, this demonstration was to verify the hypothesis that faster introduction of EVs would be achieved through a reduction of initial costs for users through various promotional measures, the use of incentives for EV usage, and improvement in services for users. Above all, the installation of charging stations was needed to encourage the widespread use of EVs. Surveys were to be conducted to identify how a town should be developed by eliminating the worry of EV battery depletion, and a user model was to be established through the gradual installation of charging stations in three areas. These surveys were to examine the current status of the installation of charging stations utilizing the current subsidy schemes, mechanisms used in other cities and other countries, the installation of charging facilities at public facilities, commercial facilities, business facilities and homes, and the installation processes and status of utilization of the charging stations. Moreover, after the gradual introduction of EV in the three areas, the demonstration was to consider service models that contribute to reducing the initial costs and maintenance costs for EV users and the improvement of user comfort and convenience. Such consideration was to include measures for checking remaining battery power and for managing vehicle conditions using navigation systems and intelligent transportation systems (ITS).

This demonstration was also to verify the hypothesis that EVs can be used as social infrastructure for power storage in a power supply system where a large number of PV systems have been installed. In the demonstration, the hypothesis was to verify that the reduction of EV's well-to-wheel CO_2 emissions and the improvement in the PV system utilization rate would be achieved and the cost to the general public could be minimized while maintaining user's satisfaction through the development of a chargeable/dischargeable EV to be utilized as electricity storage system for renewable clean energy such as PV.

Lifestyle Reforms

The following three hypotheses were to be validated in this demonstration.

(1) Changes in behavior regarding reduction of CO_2 emissions can be encouraged through the introduction of HEMS and BEMS and the visualization of energy consumption, power output, and the amount of CO_2 emissions.
(2) The citizens' environmental awareness will increase and action to reduce CO_2 emissions will be accelerated through visualization and sharing at the community level information regarding energy usage (including electricity and gas) of households through the introduction of citizen participatory SNS.
(3) Citizens will establish lifestyles that will enable the efficient utilization of energy through the introduction of economic incentives to achieve mutual supplementation between the energy supply side and the energy demand side.

5.3.3 Demonstration Results

The implementation period of the Yokohama Smart City Project was the five-year period from 2010 to 2014, and partial results of the project are beginning to be announced.[2] The demonstrations are diverse, and the task of verifying detailed results requires a certain amount of time. Nevertheless, some tangible results have already been announced in areas such as comparisons of target figures with performance figures, and the development of technology and systems (Table 5.1).

First of all, in targets for the development of a low-carbon city, results exceeded targets. At the outset of the project, the goal for CO_2 emission reductions was 30,000 tons but the actual reduction was 39,000 tons, exceeding by far the target figures. As a result, the project also exceeded the CO_2 target reduction rate of 4 %. Likewise, figures exceeding targets in the number of installations of PV and HEMS systems and the number of EVs introduced were achieved. In addition to these results, results as of the present in the area of demonstrations for CEMS, HEMS, BEMS and next-generation transportation systems are described below.

Results for CEMS

(1) Maximum reduction of 22.8 % in electric power consumption achieved through the integration of BEMS (FY 2013, figure disclosed by the City of Yokohama)
(2) Maximum reduction of 15.2 % in electric power consumption through HEMS (FY 2013, figure disclosed by the City of Yokohama)
(3) Up to 5 % improvement in forecast accuracy of electric power demand forecast (FY 2013)
(4) Successful development of SCADA (supervisory control and data acquisition), a storage battery system where a number of storage cells together can be virtually deemed to be one storage cell
(5) Successful development of a storage battery system with an interface that allows for consolidation (storage battery for home use, business use, and grid power use).

Results for HEMS (Examples of Success in Multiple Dwellings)

(1) Maximum of 12 % in energy saving, 45 % in CO_2 reduction, and 58 % reduction in peak energy demand in performance figures for FY 2013 in an initiative for total management of improvement measures for housing development, facility development and lifestyle development.

[2] Specific details of project results stated below are based on the Materials for the Next Generation Energy and Social System Meeting by the Ministry of Economy, Trade and Industry, compiled in May 2014.

Table 5.1 Results of the Yokohama Smart City Project

	Target	Results
Reduction in CO_2 emissions (tons)	30,000	39,000
Rate of CO_2 reduction (%)	25	29
Rate of reduction in peak demand (%)	20	20
Rate of energy savings (%)	17	17
Power generated through introduction of PV systems (MW)	27	36
No. of homes in which HEMS is installed	4000	4140
No. of next generation vehicles introduced	2000	2300

Source Compiled from Materials for the Next Generation Energy and Social System Meeting

Results for BEMS

(1) Development of an integrated BEMS with a function for compiling demand response distribution plans based on a negawatt transaction format and demand response substitute functions
(2) Development of a smart BEMS for optimal operation through coordination between heat storage, power generation, and battery storage
(3) Development of a multiple energy system through smart BEMS and large-scale stationary lithium ion storage battery systems.
(4) Development of a hybrid storage system that utilizes the advantages of a lithium ion capacitor and lithium ion battery
(5) Development of a BEMS that enables autonomous energy saving control based on the tenant's judgment
(6) Achievement of an overall maximum of 22 % reduction and summer maximum reduction of 22.78 % in peak demand in a demand response demonstration.

Results for Next-Generation Transport Systems

(1) Developed functions for chargeable/dischargeable PCS and EVs and communications for storing and effectively using photovoltaic power in EVs by connecting EVs with various energy management systems
(2) Developed an EV car sharing service that appropriately manages and controls storage batteries placed at charging stations, photovoltaic power generation systems, eco recharging stations that use rechargers, and EMS installation and devices
(3) Developed an integrated storage and charging system that enables multiple EVs to be recharged simultaneously in a short time.

5.4 Value Co-creation in the Yokohama Smart City Project

By what process, then, did companies that participated in the Yokohama Smart City Project achieve co-creation with other companies? As stated earlier, this project was initiated by the government and the roles of the respective companies were determined according to a master plan that had been prepared in advance. This raised the question as to whether emergence as described by Mintzberg (Mintzberg 1973, 1978, 1990) would occur in this environment. In this case, the issue was to determine whether or not new value unintended at the time of the preparation of the master plan would be created in the course of the project. It is generally accepted that emergence is apt to occur in highly autonomous "ba," whereas in the case examined here the likelihood of its occurrence is low. If emergence did not occur, it can be said that the project verified through the demonstration of value foreseen in the master plan. Even in that case, however, it would seem that some form of value would have been co-created among the companies participating in the project. I will clarify these points based on interviews I conducted.

5.4.1 Interview with Toshiba

In March 2015, I visited Toshiba and conducted interviews with staff there. The staff who cooperated in the interviews were from the Energy Solutions Development Department, the Synthesis Center of the Community Solutions Group. As a core member of the Yokohama Smart City Project, Toshiba participated in a large number of demonstrations including CEMS, BEMS and HEMS, and played a major role in these. While the interview questions were diverse, my main interest was to determine whether or not value co-creation occurred through emergence among companies in the course of executing the project. Therefore, I focused on this point during the interviews.

To state the conclusion beforehand, no phenomenon that could be described as co-creation among companies through emergence was manifested during the project. To give an example, various companies including Toshiba, Panasonic, Meidensha, TEPCO, Tokyo Gas, and Accenture participated in the demonstrations for the introduction of renewable energy in the three areas. Although I asked whether any new event occurred in the process of the demonstrations or as a result of the demonstrations, the response was that it was not possible to confirm such an event. The division of responsibilities of the respective members in the demonstrations was roughly as explained below. The three companies Toshiba, Panasonic, and Meidensha were in charge of developing various systems (HEMS, BEMS, CEMS) for measuring decentralized power sources and controlling generated output. The two companies TEPCO and Tokyo Gas provided their own buildings and company housing for installing photovoltaic (PV) systems, solar thermal conversion systems, and heat pumps. Accenture verified the effects of CO_2 reduction and economic performance assessments while the city of Yokohama promoted the dissemination and expansion of PV. According to Toshiba, the demonstrations proceeded with each company performing its role “at its own post,” and for the results of the demonstrations they made comparisons of the data they obtained with values predicted beforehand, and examined the effects. In other words, for each of the demonstrations a hypothesis was established beforehand regarding the figures and values to be obtained during the trial, and the trial was conducted to verify that hypothesis. During the demonstration process, a number of new technologies and systems were developed but in these cases too nothing was supposedly created through emergence during the demonstration process, but proceeded as intended beforehand during the planning stages.

The question, then, is did “ba” conducive to co-creation exist during the Yokohama Smart City Project? According to project information made available to the public, an organization called the YSCP Promotion Council was formed by the Yokohama Smart City Project, and within this council were joint meetings and working groups in which the respective companies participated. It was a framework for engaging in deliberations concerning how the project should proceed and the coordination of the interests of the companies. However, such “ba” supposedly existed only to confirm the sharing of roles of the respective companies, and the reconciliation of interests was done mainly by the government. In other words, there were no place for actual discussion. From the standpoint of a resource-based view, an organization’s competitive advantage is determined by the extent to which it amasses high-quality management assets within the organization (Barney 1986, 1991, 2002; Hamel and Prahalad 1990, 1994; Itami 1987; Wernerfelt 1984, 1995). For example, if the joint meetings and working groups established in the YSCP Promotion Council had roles as dynamic “ba” that promoted the exchange and integration of technology and know-how, for example, in the process of the demonstrations or based on the results of the demonstrations, new ideas and values would have been created one after another. In reality, however, in the name of confidentiality, the technologies and know-how each company possessed was considered mutually off-limits. Consequently there was no deepening of their exchanges or integration. According to Toshiba, the only area in which efforts were made by the respective companies to merge technology was in the standardization of the demand response interface. In collaboration among companies, the issue as to how open a company should be in regard to its proprietary technology and know-how is always a problem. In other words, in the name of protecting the intellectual assets of their companies, individual companies have difficulties in creating innovation (Teece 2000).

In the demonstrations in which Toshiba participated, the companies did not engage in the exchange or integration of their proprietary knowledge resources in technology or know-how at a deep level. On this point, some of the companies that participated in the demonstrations supposedly expressed the view that had there been more exchanges among companies, new activities and cooperation beyond the framework of the project would have occurred. Moreover, in projects involving a number of businesses, a company to serve as "consolidator" is necessary but in the YSCP's individual projects, there was no company that served in this capacity. Consequently, some expressed the view that the absence of such a person impeded the progress of the project. In other words, no clear leader company existed. As a result, while all of the companies worked enthusiastically at the demonstrations of technology and systems in each of the HEMS and BEMS projects, there were some companies which, compared to the HEMS and BEMS demonstrations, showed little interest in the CEMS demonstration which consolidated the other two projects. The reason for this is supposedly that CEMS needs to be consolidated among businesses and that under the current circumstances, it would be hard to promote as a business. Therefore, it is said that there was a kind of game-theoretic bargaining among the participating companies over which one would take on the volunteer-like role for consolidating (Brandenburger and Nalebuff 1996; Ghemawat 1997; McAfee 2002) CEMS.

Nevertheless, it cannot be said that no co-creation of value occurred whatsoever in the YSCP. On the basis of the interviews with Toshiba staff alone, I can at least say that it was not possible to confirm any indication of what could be called value co-creation among companies through emergence. On the other hand, it was possible to confirm certain indications of value co-creation generated through collaboration between the government and companies. One indication was in initiatives to establish citizens' cooperative organizations such as the Yokohama Smart Business Council and Energy Solutions Center Yokohama. As stated earlier, one of the objectives of the YSCP is to link results of the demonstrations to the expansion of business. While interest in smart cities is growing, there are still many issues to be resolved in its deployment as a business. For example, in the domestic market in Japan, unless legislation concerning the separation of power generation and transmission is put in place, it is difficult to establish businesses in this area. Furthermore, in areas such as CEMS explained above, where profits cannot be expected at present, the government must cooperate with companies to establish these systems as viable business areas. The Yokohama Smart Business Council and the Energy Solutions Center Yokohama are organizations which the city of Yokohama and companies participating in the YSCP are trying to establish. In the future, the government and the corporate sector must engage in initiatives to establish business while cooperating through such organizations.

Such initiatives have already begun in this project. One of these is an incentive-type demand response business. The incentive-type demand response was a demonstration conducted by Toshiba and TEPCO cooperatively. For this demonstration, reduction in electric power usage in peak demand and power saving were achieved by electric power companies which requested customers to curb their power usage and provided them with remuneration in exchange for their

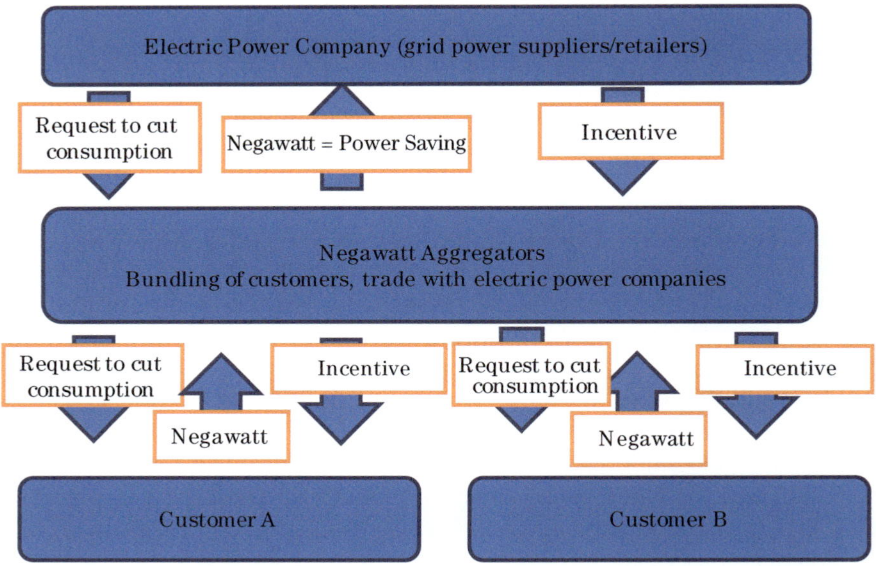

Fig. 5.2 Incentive-type demand response. *Source* Based on information provided by Toshiba

cooperation. Acting as negawatt aggregator in this demonstration, Toshiba bundled customers in groups and engaged in transactions with electric companies.[3] With the intention of turning incentive-type demand response into a business with the cooperation of electric companies and electronics manufacturers, the Yokohama Smart Business Council aims to develop a business that will optimize energy costs for customers. To do this, the city of Yokohama is also participating and the government and companies are attempting to co-create value in the electric power business through collaboration (Fig. 5.2).

5.4.2 Interview with Hitachi

In the same way as at Toshiba, I also conducted interviews with staff at Hitachi. For the interviews at Hitachi, I forwarded a questionnaire beforehand and had staff provide their responses in written form. Participating in the questionnaire interview were staff in charge of Hitachi Limited's Energy Solutions Company, Solution

[3]A concept advocated by Amory Lovins of the Rocky Mountain Institute in the United States, negawatt refers to a way of thinking that views surplus electric power achieved through energy saving by customers as being on a par with electric power generated.

Business Division. The questionnaires were forwarded in March 2015 and responses were received the following month in April.

In the YSCP, Hitachi participated in the demonstration for Next-generation Transport Systems together with Nissan Motor, Orix and Orix Auto Corporation. The objective of this demonstration was to establish an energy management system using chargeable/dischargeable EVs (EV-EMS) and to achieve reductions in peak demand and demand shifts through the combined use of renewable energy, EV storage batteries, and stationary storage batteries. In this demonstration, the roles of the companies were clearly established. Hitachi was in charge of the overall management of the EV-EMS system, Nissan Motor the development of the chargeable/dischargeable EV system, and Orix and Orix Auto Corporation the development of an EV car sharing management system. The processes of the demonstration promoted through the roles of the respective companies were as follows.

(1) Reservation information on car sharing and EV battery information was obtained from data centers.
(2) For the EV-EMS, an EV recharging schedule was drawn up taking into consideration the time of commencement of use based on PV forecast data and storage battery information within facilities.
(3) Charging was initiated to coincide with reservation conditions by dividing charging operations into two modes, a mode that would maximize use of photovoltaic power generation and a rapid mode for charging in a short period.
(4) This arrangement provided for EV charging that was both convenient for users and efficient.
(5) The system is to be connected to homes, buildings and community energy management systems.
(6) This system operates efficiently for storage batteries of stationary EVs connected to eco chargers and stationary storage batteries and improves the utilization ratio of renewable energy.
(7) This system contributes to the energy management of the community as a whole.

How, then, did Hitachi view its collaboration with various other companies during this demonstration? I endeavored to inquire about this aspect and received responses like the following.

> The four companies formed a consortium in the course of the demonstration and on a regular basis exchanged opinions and shared know-how. This cooperation was ongoing from the planning stage to development and throughout the demonstration. For example, the demonstration for the "Vehicle to Home" was originally Nissan Motor's role but Nissan went beyond that and proactively offered suggestions to all companies regarding energy management systems for the efficient use of renewable energy using EV car sharing.

I also asked about the significance of collaboration with companies of other industries for Hitachi, and received answers such as the following.

> Through collaboration with a car company, credit card company and car sharing company, we were able to share know-how that is difficult to obtain in the course of ordinary business activities, and through that we were able to gained the opportunity to be able to accumulate high-quality information across broad areas including not only system provision but also management. This is a very valuable resource for Hitachi.

What can be interpreted from the above responses is that the "ba" in the demonstration for Next-generation Transportation Systems in which Hitachi participated was not a "ba" just for the sake of formality but one that actually functioned and one where collaboration of the companies created meaningful value. At least, this is what Hitachi believed as a core member of the demonstration. The demonstration for Next-generation Transport Systems was one of seven demonstrations conducted by the YSCP, and as a demonstration for verifying a hypothesis that was fundamentally established under a master plan prepared by the government beforehand, it was analytical in nature. Despite this, the "ba" that formed was not just a formal "ba" for confirming the roles of the respective companies but was a "ba" for effectively deepening discussion and cooperation among the companies. Moreover, there were indications that there was a company that went beyond the role it was assigned beforehand to actively make suggestions. In other words, phenomena occurred that could not simply be explained through an analytical approach.

Nonaka et al. have indicated in a series of studies (Nonaka 1988, 1994; Nonaka and Takeuchi 1995; Nonaka et al. 2008; Ichijo and Nonaka 2007) on Japanese companies that a characteristic of Japanese companies is their "middle-up-down" management. Middle-up-down refers to management that is neither top-down nor bottom-up in nature, but occupies a place in the middle, making various adjustments in the top and bottom. This case comes under top-down as a project initiated under a master plan prepared by the government where each company was to execute a given role. On the other hand, the occurrence of various phenomena and presentation of ideas from the workplace level not intended in the master plan in the course of executing the demonstrations can be said to be bottom-up actions. In this context, then, middle-up-down refers to behaviors that take into account "emergence" from the workplace level in the course of faithfully executing the master plan. In specific terms, the consortium formed in the course of the demonstration performed the role of the middle, and actions such as cooperation and regular exchanges of opinions in the consortium for sharing know-how and suggestions made beyond the framework of the master plan to make the demonstration results more productive can be considered "middle-up-down" actions.

The YSCP was a project initiated by the government. How, then, did Hitachi view the role and involvement of the government in the project? Hitachi's response to this was that in the course of the demonstration the companies received various amenities from the government (Yokohama Municipal Government), and the demonstration proceeded as they engaged in discussion with the government. Furthermore, the interests of the four companies undertaking the demonstration were not always in agreement and it was necessary for the government to make adjustments on occasion to avoid any clash of interests. Hitachi also recognized that the establishment of a place for regular deliberation between the government and the four companies in the course of the demonstrations was meaningful in terms of the government's reconciling the various interests. According to Hitachi, in the YSCP the government did not demonstrate strong leadership or lead the project in a top-down manner. Of course, the project plan was initiated and prepared by the government but the government did not have strong involvement in the demonstration processes. However, the government did take charge of harmonizing the interests of the companies. This is an extremely important point. In other words, it can be said that the government performed the role of "ba" leader.

Among the four companies participating in the Next-generation Transport Systems, Hitachi and Nissan Motors are major global companies with superior technological capability. Collaboration among electronics manufacturers and carmakers is noticeable at present, and there is a strong likelihood that collaboration between companies with such outstanding dynamic capability will create innovation (Teece 2007). Hitachi has placed development of social infrastructure at the core of its growth strategy, and has expressed its expectations for further collaboration in the integration of technologies and systems.

5.5 Implications

In conclusion, I would like to summarize implications gained from observations in this chapter. As indicated several times already, the YSCP was a government-initiated project with a project budget, project period, method of implementation and division of roles that were clearly defined according to a master plan prepared by the government, and the objective of the project demonstrations was to verify hypotheses established beforehand. Therefore, for the analysis of this project, I did not believe adopting the theoretical framework of a relationship-based strategy already established in this book was appropriate. Instead, under an analytical approach, I made the assumption beforehand that it would be necessary to have a viewpoint that would consider how value co-creation would take place in this project. With an awareness of such an issue, I conducted interviews at two companies, Toshiba and Hitachi, which participated in the demonstrations as core members of the YSCP. The details of the interview are as stated above and, as noted, the responses from the two companies were quite different.

In each of the demonstrations conducted by the YSCP, as a company, Toshiba had the greatest involvement and also served as the executive head of the YSCP Promotion Council. Without a doubt, Toshiba was a company at the very core of the YSCP. According to Toshiba's view, the demonstrations conducted by the YSCP were analytical in nature, with hypotheses that were to be verified, and each company simply fulfilled its respective role "at its own post" based on a division of roles established beforehand. In Toshiba's view, the standardization of the demand response interface was the only outcome in the exchange and integration of technologies among companies, and the technology of the respective companies was mutually considered inviolable in the name of confidentiality. Moreover, according to Toshiba, the joint meetings and working groups of the YSCP Promotion Council, which were established as place for the deliberation of the companies, were actually merely formal “ba” for confirming the roles of the respective companies and did not develop into “ba” where creative views for promoting the project were exchanged. In other words, Toshiba was of the opinion that the YSCP was merely a project for verifying hypotheses assumed beforehand on the basis of a master plan prepared by the government, and that no phenomenon corresponding to value co-creation occurred among companies through emergence.

On the other hand, Hitachi’s view in this regard was quite different. Although both Hitachi and Toshiba were core members of the YSCP, Hitachi’s involvement was less than that of Toshiba. Hitachi’s actual participation was in the demonstration for the Next-generation Transport Systems only. In that sense, there is a strong possibility that Toshiba’s understanding of the circumstances of the YSCP is more accurate. In the demonstration for Next-generation Transport Systems in which Hitachi participated, however, a phenomenon that differs from Toshiba’s view clearly did take place. That is, in the demonstration where the four companies including Hitachi participated, a consortium was formed, opinions were regularly exchanged, and know-how was shared as the demonstration proceeded with the cooperation of the four members. During that process, suggestions that went beyond the framework of the roles assigned to the companies under the master plan were actively offered. As a result, Hitachi has the view that as a company it was able to accumulate know-how in broad areas and obtain valuable knowledge assets through the demonstrations.

The question, then, is how should this difference in views between Toshiba and Hitachi be understood. One fact emerges in the detailed analysis of the demonstration in which both Toshiba and Hitachi participated. That is, the difference in strategies of the participating companies. The four companies that participated in the demonstration for the Next-generation Transport Systems, particularly Hitachi and Nissan, had clear strategies to use the demonstration for creating new value for their companies. For example, in Hitachi’s case, the social infrastructure business is positioned as a pillar of the company’s growth strategy and, as an electronics manufacturer, it has high regard not only for hard aspects of the project but also general-interest infrastructure including soft aspects. Therefore, Hitachi considers know-how it has acquired from collaboration with credit card companies like Orix as a valuable asset.

In the case of Nissan Motor too, the company has high expectations for the dissemination of electric cars, and has made value creation based on electric vehicles a pillar of its business strategy. Through this demonstration, it is said that Nissan is promoting EVs not only as a car for use as a means of transportation but also for the efficient use of renewable energy including photovoltaic energy using the storage battery in the EV. Moreover, Nissan intends to promote EVs as cars that will contribute to the energy management of the local community as a whole through its use as a tool for transporting energy in times of a disaster.[4] Such ideas are clearly ideas that go beyond ideas to create conventional cars, that is, they go beyond the idea of producing cars as single items, and can be considered a clearly strategic approach intended at new value creation in EVs. In this way, the fact that the participating companies came to the demonstrations with clear strategic ideas can be considered a manifestation of phenomenon as described above.

On the other hand, in the demonstrations in which Toshiba participated, it was not possible to discern clear strategic thinking in the participating companies. As noted earlier, Toshiba collaborated with a large number of companies in the demonstrations for HEMS, BEMS and CEMS. However, in each case, the main objective was to validate hypotheses by obtaining technical data, and it is difficult to discern any other indication of strategic thinking. What becomes evident from such a review is that even in a project for verifying hypotheses as in the case of the YSCP, if the participating members have clear strategic thinking and aspire to co-creation of value among companies, actions that will lead to value co-creation can occur in the course of the demonstration processes.

An important point that can be examined when analyzing the collaboration between companies is the issue as to whether and to what extent one company will disclose to another company information regarding its knowledge assets, such as technology and know-how that it possesses. It is a well-known fact that there are companies which are very wary about engaging in active exchanges with other companies due to their overwhelming concern about the outflow of technology or know-how to other companies. However, the adoption of such an attitude is not conducive to the development of a relationship of trust among companies, which is considered a vital element in successful collaboration (Kanter 1994). Under such circumstances, generating value co-creation is difficult. Such an attitude was also evident among the companies participating in the YSCP demonstrations. As noted earlier, there was a tacit understanding among the participating companies that the intellectual resources of the respective companies were in principle an inviolable area under the pretext of confidentiality. In research on the subject of open innovation that has a significant impact on collaboration and value creation among companies, several scholars have indeed indicated the importance of creating innovation through collaboration with other companies where those companies disclose proprietary technology and know-how in their possession (Chesbrough 2003a, b, 2006, 2007; Kirschbaum 2005; Grassmann 2006). However, such openness does not mean revealing everything to the other parties. It goes without saying that there are technologies and know-how that a company does not make available to other companies. In other words, an open innovation strategy is a strategy that in strict terms means both open and closed innovation. The issue for a company is to determine which technologies and know-how it will reveal and which technologies and know-how it will maintain as confidential. In other words,

[4]During its participation in the YSCP demonstrations, this view of Nissan's was expressed repeatedly by Nissan stakeholders.

from the perspective of value creation, it is vital for a company to have the capability to accurately judge which of its intellectual assets to reveal and which to keep confidential. When companies with this capability come together and engage in collaboration, the respective technologies and know-how which they have will resonate and result in the co-creation of value.

References

Andrews, K. R. (1971). *The concept of corporate strategy*. Homewood, Ill: Dow Jones-Irwin.

Ansoff, H. I. (1965). *Corporate strategy: An analytical approach to business policy for growth and expansion*. New York: McGraw-Hill.

Barney, J. (1986). Strategic factor markets: Expectations, luck, and business strategy. *Management Science, 32*(10), 1231–1241.

Barney, J. (1991). Firm resources and sustained competitive advantage. *Journal of Management, 17*(1), 99–120.

Barney, J. (2002). *Gaining and sustaining competitive advantage* (2nd ed.). Upper Saddle River, NJ: Prentice Hall.

Brandenburger, A. M., & Nalebuff, B. J. (1996). *Co-opetition*. New York, NY: Currency.

Campbell, T. (2012). *Beyond smart cities: How cities network, learn and innovative*. New York: Routledge.

Chesbrough, H. (2003a). The era of open innovation. *Sloan Management Review, 44*(3), 35–41.

Chesbrough, H. (2003b). Open innovation: How companies actually do it. *Harvard Business Review, 81*(7), 12–14.

Chesbrough, H. (2006). *Open business models: How to thrive in the new innovation landscape*. Boston, MA: Harvard Business School Press.

Chesbrough, H. (2007). The market for innovation: Implications for corporate strategy. *California Management Review, 49*(3), 45–66.

Deakin, M. (Ed.). (2014). *Smart cities: Governing, modeling and analyzing the transition*. New York: Routledge.

Ghemawat, P. (1997). *Games business play: Cases and models*. Cambridge, MA: The MIT Press.

Grassmann, O. (2006). Opening up the innovation process: Towards an agenda. *R&D Management, 36*(3), 223–226.

Hamel, G., & Prahalad, C. K. (1990). The core competence of the corporation. *Harvard Business Review, 68*(3), 79–91.

Hamel, G., & Prahalad, C. K. (1994). *Competing for the future*. Boston, MA: Harvard Business School Press.

Hollands, R. (2008). Will the real smart city stand up? *City, 12*(3), 302–320.

Ichijo, K., & Nonaka, I. (Eds.). (2007). *Knowledge creation and management: New challenges for managers*. Oxford: Oxford University Press.

Itami, H. (1987). *Mobilizing invisible assets*. Cambridge, MA: Harvard University Press.

Kanter, R. (1994). Collaborative advantage: Successful partnerships manage the relationship, not just the deal. *Harvard Business Review, 72*(4), 96–108.

Kirschbaum, R. (2005). *Open innovation in practice. Research-technology management, 48*(4), 24–28.

McAfee, R. P. (2002). *Competitive solutions: The strategist's toolkit*. Princeton, NJ: Princeton University Press.

Mintzberg, H. (1973). Strategy-making in three models. *California Management Review, 16*(2), 44–53.

Mintzberg, H. (1978). Patterns in strategy formulation. *Management Science, 24*(9), 934–948.

Mintzberg, H. (1990). The design school: Reconsidering the basic premises of strategic management. *Strategic Management Journal, 11*, 171–195.

Nonaka, I. (1988). Toward middle-up-down management: Accelerating information creation. *Sloan Management Review, 29*(3), 57–73.

Nonaka, I. (1994). Dynamic theory of organizational knowledge creation. *Organization Science, 5* (1), 14–37.

Nonaka, I., & Takeuchi, H. (1995). *The knowledge creating company*. New York: Oxford University Press.

Nonaka, I., Toyama, R., & Hirata, T. (2008). *Managing flow: A process theory of the knowledge-based firm*. London: Palgrave macmillan.

Steiner, G. A. (1969). *Top management planning*. New York: Macmillan.

Teece, D. J. (2000). *Managing intellectual capital: Organizational, strategic, and policy dimensions*. New York: Oxford University Press.

Teece, D. J. (2007). Explicating dynamic capabilities: The nature and microfoundations of (sustainable) enterprise performance. *Strategic Management Journal, 28*(13), 1319–1350.

Wernerfelt, B. (1984). A resource-based view of the firm. *Strategic Management Journal, 5*, 171–180.

Wernerfelt, B. (1995). A resource-based view of the firm: Ten years after. *Strategic Management Journal, 16*, 171–174.

Chapter 6
Theoretical and Managerial Implications

6.1 Introduction

In this chapter, I will consolidate the various implications drawn from the three case studies on smart city projects examined in Chaps. 3–5, and verify various academic indications regarding the establishment of competitive advantage through co-creation, which is the main theme of this book. In each of the chapters above, efforts were made to draw out implications from the case studies presented. The task in this chapter is to draw out implications with more universality by juxtaposing and comparing the implications of the individual case studies. In theoretical structures in research on organizations, "disciplined imagination" is considered an important element (Weick 1989, 1995). Applying disciplined imagination here will require clarifying the explanatory variables and explained variables of the respective three cases and comparing their causal relationships rather than simply listing the various implications drawn from the three case studies as projects conducted in different environments under different conditions. The implications identified through this process will theoretically provide more in-depth clues to understanding the establishment of competitive advantage through co-creation, the overarching theme of this book.

6.2 Implications Drawn from the Three Case Studies

Each of the three case studies taken up in this book was different in nature. For example, the case of the Yokohama Smart Community discussed in Chap. 3 was a project where the autonomy of the participating companies was respected on one hand but the presence of a leader to take on the role of coordinator of the member companies of different sectors was not clear. In contrast to this was the case of the

© Springer Japan 2016

N. Tokoro, *The Smart City and the Co-creation of Value*,
SpringerBriefs in Business, DOI 10.1007/978-4-431-55846-0_6

Fujisawa Sustainable Smart Town (Fujisawa SST) examined in Chap. 4, which had a clearly defined leader (Panasonic) and was a project conducted at the initiative of the leader. In the third case in Chap. 5, the roles of the participating companies were clearly defined, and their autonomy was limited under a government-led demonstration project. In analyzing these projects with such different natures as described above, I adopted a theoretical framework for relationship-based strategies consisting of the three components "ba," emergence, and synthesis, and conducted analyses to obtain promising insights into the establishment of competitive advantages achieved through the co-creation of companies of different sectors. From the perspective of the three vantage points of (1) "ba" and emergence, (2) the existence of a leader and (3) analysis and synthesis, I would now like to verify implications drawn thus far.

6.2.1 "Ba" and Emergence

"Ba," or shared context in motion, is the key concept in relationship-based strategies. Furthermore, the existence of "ba" is essential for companies of different sectors to interact and create new value. In other words, it can be assumed that between the nature of the "ba" and value co-creation, a certain causal relationship exists. This book establishes the hypothesis that the more autonomous the "ba," the more apt interaction and integration of knowledge and know-how among the participants are to occur, and the greater the possibility value co-creation will take place. Underlying the establishment of this hypothesis is the research of numerous scholars (Bower and Gilbert 2007; Mintzberg 1973, 1978, 1990; Mintzberg and Waters 1985; Quinn 1978, 1980; Burgelman 1983, 1994, 2002) and the concept of knowledge creation theory (Nonaka and Takeuchi 1995; Nonaka and Konno 1998; Nonaka et al. 2008). All of these scholars share the view that there is a strong likelihood an autonomous "ba" induces emergence and knowledge creation.

Of the three case studies in this book, the "ba" with the most autonomy was the Yokohama Smart Community project. In the Yokohama Smart Community, the participating companies formed a consortium as their "ba," which promoted the project through a mechanism whereby the participating companies brought their respective technologies and know-how to the "ba." Moreover, the consortium was extremely open in nature and had a high level of freedom. Almost no restrictions were placed on the participating companies when they participated in the consortium. Moreover, the consortium had neither enrollment fees nor compulsory dues for administration of the consortium. In addition, the consortium had neither a budget nor plans that had been established beforehand. The operational principles of the consortium and decisions whether or not to proceed with a project were all decided in mutual discussion among the companies participating in the consortium. Each project was managed by the participating companies which brought to the project their proprietary technologies and know-how. Costs required for the project were settled after the project was completed. Moreover, the consortium itself owned no intellectual property rights, and there was an understanding that any intellectual property rights including rights to technology and know-how developed through the project belonged to the participating companies.

It is believed that a “ba” like this with a high degree of freedom is highly likely to generate emergence and knowledge creation, and knowledge generally recognized as corresponding to this was verified in the case analysis of the Yokohama Smart Community. In this case, a project that commenced with the Fukuoka Smart House Consortium continually expanded its circle of activities from one to the next, gradually changing the participating members and project content, and these activities led to the launch of the Yokohama Smart Community and Nagasaki Smart Society. These projects had not been planned beforehand, but evolved naturally from ideas that arose in the course of pursuing the original project. All of the members participating in the consortium joined as members that shared the vision espoused by the project, that is, to “establish an energy system learned from a view of life found in nature,” and not because any kind of superior-subordinate relationship such as a capital relationship, or corporate affiliation existed among them. In the course of free discussion in the consortium, ideas for new projects, or “emergence,” arose. The technologies and know-how individual companies brought to the projects merged and became integrated, resulting in the development of new technologies and systems.

In contrast to this case was the Yokohama Smart City Project. This was a project initiated by the government and based on a master plan prepared by the government. Each segment of the project had its own budget and implementation period, and the role each participating company was to play was clearly defined. Therefore, a “ba” like the YSCP Promotion Council, which was set up by the project, was not a body for ensuring the autonomy of the participating members but was a “restrictive ba” merely for formally confirming the roles of the respective companies participating in the project. Moreover, as Toshiba, one of the core members of the project, clearly explained during an interview, the objectives of the respective demonstrations conducted in the Yokohama Smart City Project were to validate a hypothesis established beforehand, and to compare the hypothesis and demonstration results and analyze any differences. Therefore, according to Toshiba, the origination of new activities or ideas unintended from the outset did not arise in the process of implementing the project. Although I had anticipated this prior to conducting the case analysis, I had not eliminated the possibility of emergence occurring as a possibility. This was because Mintzberg (1978) had argued that emergence was a process whereby the intended strategy divided into an unrealized strategy and a deliberate strategy as time passed, and an emergent strategy that appeared amid these combined with the intended strategy to become the actually realized strategy. This essentially suggests that even when a plan that is established in a top-down manner beforehand is implemented, there is a possibility that emergence will occur in the implementation process. Nonaka and Takeuchi (1995) had also indicated that the tendency for emergence to occur was a characteristic of organizations in Japanese companies.

Moreover, assertions like those of Mintzberg and Nonaka et al. had already been confirmed to some extent in the case analysis of the Yokohama Smart City Project. This was in the demonstration for Next-generation Transport Systems led by Hitachi and Nissan. In this demonstration, actions that clearly led to emergence among participating members occurred. Even in an environment where the roles of the respective companies are clearly defined and hypotheses to be validated are established beforehand under a master plan, new actions beyond those planned beforehand can occur if the participating members share a vision to create new value by promoting mutual exchanges and integration of knowledge and know-how.

Results of this analysis clearly indicated that a certain positive correlation existed between a "ba" high in autonomy and emergence but also prompted the implication that even if the autonomy of a "ba" were restricted, the awareness of a purpose among the members of the "ba" could give rise to emergence.

6.2.2 Presence of a Leader

Analysis of the case study of the Yokohama Smart Community confirmed that the greater the autonomy of the "ba," the more apt emergence is to occur among members participating in the "ba." On the other hand in "ba" where the autonomy of participating members is respected to an excessive degree, the problem of the absence of a "ba" leader, that is, a mediator or organizer, arises. This was indeed a problem in the Yokohama Smart Community consortium. As noted earlier, while the consortium of the Yokohama Smart Community was a "ba" that was open, had a high level of freedom, and respected the autonomy of the participating members, there was no distinct presence of a leader presiding over the "ba." Although core companies did exist, there was no demonstration of the kind of initiative that encourages co-creation among companies of different sectors. Renowned for his research in "ba" management, Itami (1999) argued that the presence of a leader to appropriately manage the "ba" was essential for the invigoration of the "ba," creative interaction among members participating in the "ba," and the creation of value.

This gives rise to the question as to what disadvantages occur when the presence of a "ba" leader is lacking. One scenario that can be imagined is the clash of interests among members. In other words, if members of various backgrounds are participating in a "ba" and appropriate coordination of interests is lacking, the "ba" will become a stage where clashes in interest among members take place repeatedly. To avoid such a situation, a "ba" leader to coordinate the various interests of members is essential. In the case of the Yokohama Smart Community, no serious clashes of interests among members in the consortium surfaced. Despite the absence of a clear "ba" leader, core companies serving at the core of the consortium did exist, and there is a possibility that these companies fulfilled the implicit role of coordinator of interests. However, this was a project which in principle operated on the basis of the voluntary participation of members which shared the vision espoused by the project and, in the absence of a clear leader, there was a possibility that the problem of clashes in interest would surface in the course of implementing various projects.

Furthermore, the absence of a "ba" leader is also a negative factor in value creation. Without a doubt, there is a possibility emergence will occur again and again through the interaction of members in a highly autonomous "ba. " This does not mean, however, that every emergence will result in the creation of value, since emergence and value creation are not one and the same. Even when various emergences occur through the interaction of members, among the emergences that arise are those emergences that pertain to improvement in operations to some extent and do not necessarily result in overall value creation. It is the "ba" leader that selects from among the various emergences the ones that will lead to value creation, or combines various emergences to bring them to fruition in the form of value creation. Because of the absence of a clear "ba" leader to assume such a role in the case of the Yokohama Smart Community, it was difficult to determine what kind of value the activities taking place in the consortium would create.

On the other hand, Panasonic was a clear "ba" leader in the Fujisawa SST, and appropriately fulfilled that role. In other words, in the Fujisawa SST, Panasonic clearly indicated its intention to create value in the five areas of energy, security, mobility, healthcare and community, and at the initiative of Panasonic a structure for creating value in these areas through mutual interaction of the participating companies was established. In the "ba" created in the Fujisawa SST, the autonomy of the participating members was not given the same level of consideration as in the Yokohama Smart Community consortium. Fujisawa SST was basically a Panasonic project, and strongly reflected the intentions of Panasonic in every process from the drafting of the project plan through the implementation of the project. Furthermore, the other participating members were required to conduct their activities in line with those intentions. In that sense, it could be said that the consortium was a restrictive "ba," and was not an environment particularly conducive to emergence. On the other hand, it would be incorrect to say that there was no leeway whatsoever for the other participating members to demonstrate their autonomy. While the framework of the basic strategy created by Panasonic had constraints, within that framework a "ba" that actively encouraged emergence through the interaction of the members was established. As a result, new value and ideas were created through the resonance of technology, know-how and other knowledge of the participating members, and it was the role of Panasonic as the "ba" leader to further enhance the development of the five values mentioned earlier and exercise management that would achieve this.

In contrast to the general view that it is difficult to predict value creation and innovation beforehand, Christensen and Raynor (2003) argue that it is possible to predict these processes. For example, as illustrated by the case study of the Fujisawa SST, it is quite possible to predict the processes of value creation and innovation when there is a clear presence of a leader that can appropriately exercise management of the "ba." Conversely, when a "ba" lacks such a presence, it can be said that it is difficult to predict such processes.

6.2.3 Analysis and Synthesis

As already stated in Chapter 2 in this book, the relationship of analysis and synthesis originally meant opposite actions. For example, analysis is an approach to understanding a complex system by understanding that system as a whole. This is done by reducing the system to individual elements that comprise the system and making efforts to grasp the nature of the individual elements and the relationships between them. In contrast, synthesis is a method of integrating individual elements in order to create entire systems. Natural sciences, which developed to shed light on mechanisms in the natural world, have achieved success by adopting an analytical approach. For example, to understand water as a substance, the materials that comprise water were investigated and broken down into individual substances of hydrogen and oxygen. This process elucidated the fact that water could be made by combining two parts hydrogen and one part oxygen. The elucidation of this fact was reached through an analytical approach. As this example shows, the scientific approach originally meant an analytical approach, and in the world of natural sciences, the analytical approach has developed through its application it to every phenomenon from the origin of the universe to mechanisms in the lives of microbes.

On the other hand, the essence of the action of co-creation, which is the main theme of this book, can be understood by adopting a synthetic approach. This is because the creative act of co-creation is not an act of reducing a given system to grasp the nature of the respective elements or the relationship between them, but an act of creating a new system through the integration of individual elements. In the act of co-creation, however, the question arises as to whether the processes of analysis and synthesis should be clearly demarcated. Judging from the case studies I have examined in this book, I would say no. We should not consider that co-creation is essentially a process that should be analyzed through a synthesis approach, and an analytical approach should be viewed as an antithesis to this. Therefore, adopting an approach that integrates both can be considered appropriate.

The case of the Fujisawa SST provided beneficial implications in validating this point of view. Fujisawa SST is an integrated formation of analysis and synthesis, which attempts to generate co-creation in a fine balance of both. As stated earlier, the Fujisawa SST project is a project initiated by Panasonic, and strongly reflects the intentions of Panasonic from the drafting of the project plan through the implementation of the project and the value to be created. In other words, this project was basically a top-down plan analytical in nature, conceived of beforehand and executed systematically and rationally by Panasonic which enlisted the cooperation of other participating companies after predicting the value to be created.

This pattern, where the leading agent drafts the entire plan, coordinates the activities of the organization's members in line with that plan, and systematically executes the plan, coincides with the views of scholars of strategic planning (Ansoff 1965; Andrews 1971; Steiner 1969).

However, the Fujisawa SST was not a project that operated solely on the basis of a top-down, analytical approach led by Panasonic. Aiming for the creation of new value not planned at the outset of the project, Panasonic employed various tactics to draw out the knowledge of other participating companies which they exchanged with each other and integrated. A typical example of this, as described in Chap. 4, is the kind of appealing activities SO TWO offered at the Shonan T-SITE, a commercial facility in Fujisawa SST. It was not the intention of Panasonic as leader of the Fujisawa SST to generate solely on its own the value to be created at the Fujisawa SST; it aimed for creation through co-creation with other companies, and therefore established a number of "ba" for companies to engage in discussion. This was an indication that Panasonic recognized the importance of building communications networks within organizations in order to generate new ideas and values in organizations (Lazer and Friedman 2007).

In the Fujisawa SST project, a top-down approach and a bottom-up approach with Panasonic at the center were skillfully combined, and very interesting processes of value co-creation are occurring. Nevertheless, this is a project currently underway, and is set to be completed in 2018. Since the project is still at a developmental stage, clear phenomena that could be called value co-creation based on synthesis have not as of the present become manifest. Nevertheless, in view of the nature of the Fujisawa SST project, that is, the establishment of "ba" that encourage co-creation with other companies and the presence of a leader company, it is likely that value co-creation based on synthesis will occur with certainty in the future. To be correct, this means value co-creation through the integration of analysis and synthesis (Fig. 6.1).

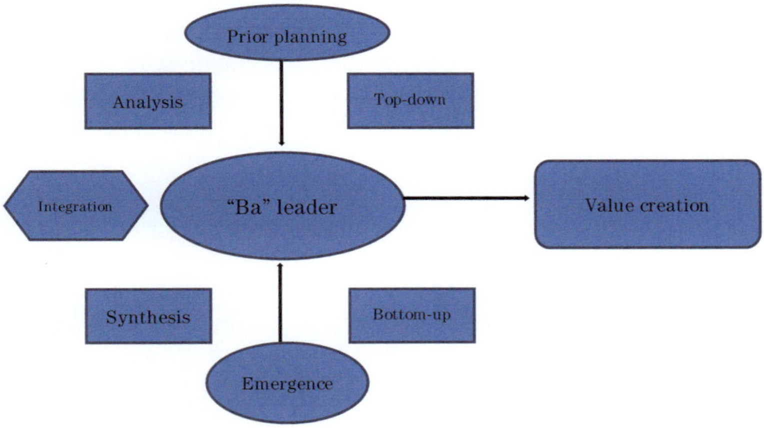

Fig. 6.1 Structure of value creation. *Source* Prepared by the author

6.3 Economic Value and Social Value

6.3.1 Relationship of Economic Value and Social Value

This book views the issue of value co-creation by companies of different sectors as a central theme but what kind of value do we assume value created through co-creation to be? Here I would like to take a moment to clarify the specific content of value. Value in corporate activities generally refers to economic value in the majority of cases and essentially means indicators that signify economic performance such as sales, net income, ROE and share price. If the figures indicating economic performance had risen, for example, if sales in a company rose 20 % year on year, or share value increased by 2 %, corporate value would be considered to have risen. When investors choose companies to invest in, these are the kinds of economic values on which they base their judgment. Corporate activities are activities basically aimed at increasing economic value, and it is impossible to engage in a discussion concerning details of value without mentioning this value. Therefore, the content of value to be created even in value co-creation among companies of different sectors, which is the central theme of this book, must include indicators of economic value such as those mentioned above. Without mention of these, such discussion is likely to become abstract and divergent.

In this book, however, there are no in-depth analyses of value co-creation and economic value. Therefore, there is no indication of specific analysis results regarding the kind of economic value companies that participated in the smart city construction projects created as a result of their co-creation of value with companies of other industrial sectors. There are two reasons for this. The first is that the main concern of this book is not to indicate specific economic value but to focus on the analysis of processes up to that stage. In other words, as in precedent research analyzing collaboration between organizations (Vangen and Huxham 2003; Kanter 1994; Dacin et al. 1997; Das and Teng 1998; Gray 1985; Carley and Christie 1992; Wistow and Hardy 1991), in this book the process analysis of value co-creation is the main focus of interest rather than how much sales increased or by what percentage the share price rose as a result of value co-creation achieved in the smart city project. The second reason is that it would have been difficult to approximate economic value at present. Although the smart city projects taken up in this book—the YSCP at the demonstration stage and the other two projects—have been achieving positive outcomes in areas such as development of technologies and system development, these outcomes are still not at the stage where they translate into economic value.

Therefore, this issue must be the subject of in-depth analyses in future research of the demonstrations. As smart cities make the transition from the early developmental stage to the growth stage and various values created through co-creation among companies of different sectors result in economic value in areas such as technologies, systems, and services, it will be necessary to clarify the relationship between co-creation of value and economic value.

Of equal importance with economic value is social value. Although social value has already been discussed in Chaps. 1 and 2, this book has refrained from making assumptions concerning economic value alone in regard to the content of value created as a result of value co-creation. Economic value is undeniably one of the most important values created in corporate activities but it is not the only value, and must not be the only value resulting from corporate activities. Corporate activities focused solely on the creation of economic value deteriorate into a state of irresponsible capitalism and consequently bring harm on society. There is not enough time here to mention the myriad cases to date that fall into this category. Even in the case of smart cities, while the original objective is to create social value in areas such as ensuring a safe, secure lifestyle and realizing a low-carbon society, it is conceivable that some companies participating in the construction and operation of these communities will give priority to the creation of economic value alone and focus their attention solely on their own companies' sales and boosting profits without paying any attention to the creation of social value. Although many companies espouse the creation of social value as one of their fundamental principles, their espousal of this ideal is all too often disregarded in their everyday activities.

In this environment, the demonstration of a growing affinity towards the concept of creating shared value (CSV) advocated by Porter et al. (Porter and Mark 2011) is a manifestation that society demands that companies create not only economic value but also social value. Porter is a world authority in the field of management strategy and has made many notable achievements as the progenitor of the "positioning view" (Porter 1980, 1985, 1991, 1996). The positioning view focuses on the acquisition of an advantageous position in the market that will afford a company the opportunity to gain a competitive advantage. In his early stages of advocating a positioning view, Porter had a decidedly biased view of corporate value as economic value. His view subsequently evolved to one more reflective of an awareness of social values and eventually arrived at a view of creating shared value (CSV).

According to Porter, the most effective approach to solving the various problems that society faces including environmental problems is to apply business methods. This is because companies have access to the most abundant resources in terms of money, technology, know-how, and human resources for solving various social problems. Applying business methods to resolve various social problems, companies not only solves problems but also enables companies to generate profits and establish a competitive advantage in the process. This is the essence of the concept of CSV.

In the concept of CSV advocated by Porter, however, there is no clear indication of the nature of the relationship between economic value and social value. Although Porter does acknowledge that corporate value includes social value as well as economic value, he does not give a clear theoretical explanation of how the two values relate to each other. In traditional corporate theory, social value is viewed as being one part of economic value, and the need to pursue social value is recognized so long as results in an increase in economic value. Friedman (1962) is a representative polemicist of this view. Furthermore, the basic understanding regarding

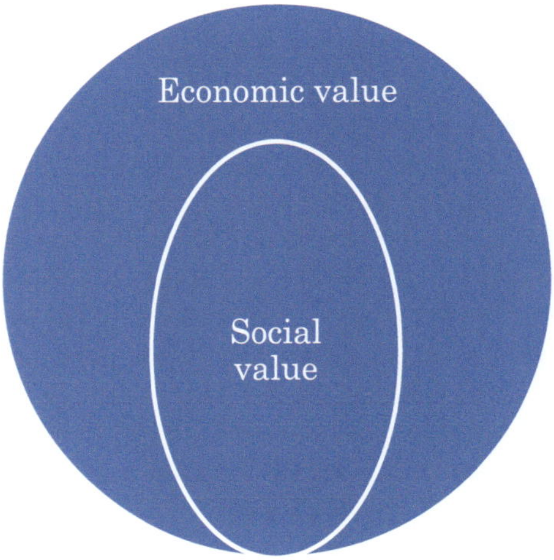

Fig. 6.2 Economic value and social value in traditional corporate theory. *Source* Prepared by the author

strategic philanthropy and initiatives focused on corporate social value such as CSR is that placing importance on such social value ultimately results in boosting the economic value of a company. In other words, in various concepts to date, the ultimate goal of corporate value has consistently been economic value, while social value has been viewed as a principle to be pursued only on the premise that it contributes to augmenting economic value. In other words, social value that fails to contribute to enhancing economic value is not worth pursuing for a company. The CSV that Porter advocated did not clarify whether the relationship of the two values adhered to the view of traditional corporate theory or whether signaled a new relationship (Fig. 6.2).

6.3.2 Integration of Economic Value and Social Value

This book presents a different view of the relationship of economic value and social value from that indicated in traditional corporate theory. Value creation through co-creation, which is the key concept in this book, includes both economic value and social value but the relationship between the two is not one of control and subordination, which traditional corporate theory assumes. The creation of value through co-creation in this book is not based on the assumption of a relationship where social value is governed by economic value, or where the existence of social value has no significance unless it contributes to enhancing economic value.

In this book, economic value and social value are viewed as having separate, independent values, and are recognized as entities that have a mutual influence on each other. In other words, social value is not a part of economic value but a value to be pursued in its own right as an independent value uninfluenced by trends in

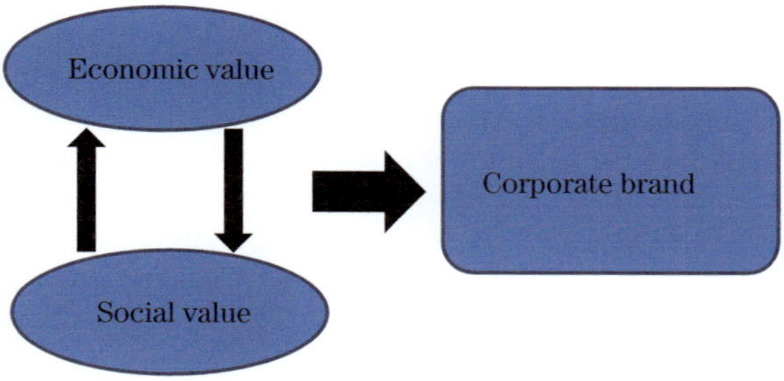

Fig. 6.3 Establishing corporate brand through the integration of economic value and social value. *Source* Prepared by the author

economic value. Furthermore, in the process of a company's pursuit of these two values, the values have a mutual influence on each other and new value is created as they merge.[1] This is the essence of value creation through co-creation that this book envisages (Fig. 6.3).

The social value to be created through the construction of smart cities will be the realization of a low-carbon society, a safe and secure lifestyle, and good health and culture, among others, and the economic value, on the other hand, will be in gains such as sales, net profit and a rise in share price. If both values are pursued as separate, independent values, the situation of most concern is one where a trade-off relationship eventuates between the two. In such a scenario, the pursuit of social value will result in a decline in economic value or, conversely, social value will be sacrificed for the sake of pursuit of economic value. For example, as described in one of the case studies taken up in this book, CEMS is an energy management system for an entire community. If the establishment of a CEMS is realized, meaningful social value will be created with certainty. On the other hand, if the CEMS is viewed as a business enterprise, it will be difficult to consider it as a profit-making business as of the present. Consequently, so long as economic value and social value are in a trade-off relationship, it will be difficult to pursue both simultaneously. Therefore, my argument here is predicated on the assumption that the problem posed by this trade-off relationship is resolved. In such a situation, the role of the government will be crucial. That is, if the creation of social value is viewed as having a significant impact on society, the government should reinforce its policy assistance in efforts to resolve the trade-off.

Once the trade-off relationship between the two values is resolved, the proposition of this book will become more plausible. The mutually positive impact that economic value and social value have on each other will then create a positive cycle, triggering synergistic results and generating further value. Consider, for example, the scenario of a housing developer providing smart houses, which are part of the smart city infrastructure. Consumers will show their support for the social value of homes with low environmental impact, sales of the houses will

---

[1]Masahiro Okada refers to new value derived from the integration of economic value and social value as "integrated value." For details, see Harvard Business Review, DIAMOND, January 2015, pp. 40–53.

increase, the market will view this trend as a positive sign, and stock prices will rise. As a result, the financial soundness of the company will increase, and this will result in further development of homes focused on social value. Within this positive cycle, economic value and social value merge, and new corporate value is created. This new corporate value encompasses both economic value and social value, and the company not only achieves high economic performance but also earns a high level of trust from society. This corporate value, which is underpinned by sound economic performance and the backing of society, is essentially what corporate brand is all about. In other words, new corporate value is born in the form of corporate brand through the integration of economic value and social value.

6.4 Building Competitive Advantage Through Co-Creation

Bearing in mind considerations up until now, I would like to consolidate the kinds of implications that have been drawn in relation to the overarching theme of this book: establishing competitive advantage through co-creation.

6.4.1 Verification of the Establishment of Two Competitive Advantages

This book has indicated two possibilities regarding the establishment of competitive advantage through co-creation. The first is the establishment of a competitive advantage through the integration of value creation through co-creation with existing corporate value, and the second is the competitive advantage achieved through the accumulation of tacit knowledge within the organization, resulting from progress in the integration of knowledge and know-how through co-creation.

In the case of the former, it is assumed that co-creation occurs among companies of different sectors during the building of the smart cities, and new value is created as knowledge and know-how are exchanged and integrated. Furthermore, it is assumed that a competitive advantage is established through integration of this new value with the corporate value of individual companies. In such case, the new value created through the exchange and integration of knowledge and know-how among companies means social value for the most part. In other words, this refers to the technologies and systems required to establish a low-carbon society, or the services required for a safe and secure lifestyle or to lead a healthy life enriched with cultural amenities. At the same time, the corporate value possessed by individual companies encompasses both economic value and social value. Economic value such as sales, profits, and share prices accounts for the major share of corporate value and is representative of corporate value. In addition to this, however, corporate value includes the social value that a given company has pursued to date. For example, this corresponds to activities such as philanthropic activities that a company has continued over many years. If new social value generated through co-creation with the existing corporate value of that company triggers a chemical reaction and new value of a higher level is created through both economic value and social value, this value will become a new competitive advantage of the company. This is the first hypothesis envisaged in this book in regard to establishing a competitive advantage through co-creation.

In this book, I examined three smart city projects that were different in nature and endeavored to validate the first hypothesis through case analyses. It is my view that the hypothesis is highly valid, but unfortunately I was unable to adequately verify the correctness of the hypothesis in the case analyses. In the three smart city projects analyzed in this book, there is no doubt that various new technologies, systems and services were developed one after the other in succession, and that new social value was created through co-creation brought about by companies hailing from different sectors. However, to determine how this social value became integrated with the existing corporate value of the respective companies and how this in turn led to a competitive advantage will require further observation on a somewhat longer time axis. In particular, the impact newly generated social value has on economic value in terms of sales, profit and share prices will again be validated when the smart city projects are established as business enterprises. However, the ultimate goal of competitive advantage envisioned in this book is the establishment of a corporate brand, not just a boost in economic value. In other words, when the corporate brand of a given company in the market is strengthened through its construction of a smart city, that will be considered the establishment of a competitive advantage.

Let us now consider the second competitive advantage. The efforts of smart city projects are promoted by companies of different sectors that bring to the respective projects their proprietary technologies, services, and know-how. This means that various technologies, services and know-how interact and merge. Through experience of this nature, the companies participating in the projects acquire valuable assets that they could not have acquired in projects conducted singlehandedly on their own, and experiences of this nature accumulate as tacit knowledge within these companies over time. When tacit knowledge that has accumulated demonstrates its validity through the development of new products and services, it is recognized as a new competitive advantage. This is the second hypothesis envisaged in this book in regard to establishing competitive advantage through co-creation.

As I stated in an earlier chapter, I conducted interviews with core companies of the various projects during the case analyses in this book. Almost all of the companies that responded to the interview replied in the affirmative to the interviewer's question as to whether they had obtained and amassed within their organizations valuable information through their cooperation with companies of different sectors.

Likewise, almost all of those interviewed answered in the affirmative to the question as to whether such intellectual assets would contribute to their companies' competitive advantage in the future. In other words, the accuracy of the second hypothesis was tentatively confirmed by the core companies of the project.

However, there was one major issue in the validation of this hypothesis. This was the fact that it was impossible to get a complete picture of the kind of intellectual assets which were acquired and amassed within the organizations of companies. While acknowledging that they acquired intellectual assets in the course of the projects, even the companies that responded to the interviews were unable to give specific details of the content of those assets. Providing a response to such a question admittedly impinges upon corporate confidentiality and at the same time can be an indication that the companies themselves may not grasp the overall picture regarding these assets. Tacit knowledge is an intangible intellectual asset and one that is difficult not only to quantify but also to describe in words. This aspect of tacit knowledge has been cited as an issue in knowledge theory, and tacit knowledge is an attribute that comes under fire as lacking in empirical proof (Eisenhardt and Santos 2002). Therefore, even if intellectual assets are acquired through co-creation between companies of different sectors and are amassed as tacit knowledge within their organizations, it is difficult to say with accuracy in what form this knowledge leads to the establishment of a competitive advantage. In the course of ongoing diligent collection of case studies, step-by-step efforts will be required to shed light on this relationship.

What can be said of verification of the establishment of the two types of competitive advantage is that some aspects of competitive advantage can be objectively grasped through quantification and visualization but others not, and they defy explanation through simplification of the causal relationship like a mathematical formula. Competitive advantage through co-creation, which has been the focus of this book, is a form of competitive advantage that is difficult to grasp objectively through quantification or visualization and, therefore, implications drawn through validation efforts have not been demonstrated in a clear form of competitive advantage. In the validation of such competitive advantage, it will be important to continue the step-by-step processes of gathering a large number of case studies through a filter that identifies "the establishment of competitive advantage gained through co-creation," collating facts that come to the fore in relation to this subject, and identifying common essential elements.

6.4.2 Requirements for Establishing "Competitive Advantage Through Co-Creation"

This book established a theoretical framework for a relationship-based strategy consisting of three components: "ba," emergence and synthesis, and considered the theme of establishing a "competitive advantage through co-creation" under this theoretical framework. The various implications drawn from the case studies of three smart city projects are as stated above but in this section I would like to probe further into these implications in efforts to draw out deeper, more universal implications.

First of all, there is the "ba," which is a key concept in a relationship-based strategy, and the existence of "ba" is an essential element for establishing a "competitive advantage through co-creation." As already mentioned regarding the concept of "ba," it is a concept that has been used in various academic fields such as physics, psychology and philosophy. "Ba" as a concept has also been applied in the area of management by Japanese scholars of business management such as Nonaka and Itami. Moreover, underlying the increasing interest in knowledge management in Europe and the United States is the widespread penetration of the concept of "communities of practice" (Lave and Wenger 1991; Wenger 1998; Wenger and Snyder 2000). Whatever the case, for new knowledge to be created through the exchange of diverse knowledge that people possess and the processes of exchanging it, the establishment of a type of space that will make this possible is essential. Such spaces may be intentionally created in an organized manner but they may also develop naturally as a matter of course.

What is important is that this space fulfills the requirements of encouraging the exchange of knowledge and enabling the creation of new knowledge. This is essentially the nature of "ba." The following two points can be made in regard to the nature of "ba." The first is clarification of the vision that expresses the objectives of the "ba," and the second is the sharing of the vision by the participating members of the "ba." Without the existence of a clear vision in a "ba," the objectives and direction of the "ba" will remain uncertain. Consequently, there is a strong likelihood the participating members will become aimless, and an unruly space will be created. There are occasions when energy for exchanging knowledge is generated from such a formless, chaotic state and new value is created, but such cases are rare. In short, a "ba" with a clear vision is an essential requirement for promoting co-creation among members participating in the "ba." This point became clear in the case study of the Yokohama Smart Community taken up in Chap. 3. In the Yokohama Smart Community, a "ba" with a clear vision to "establish an energy system learned from a view of life found in nature" was created. As a result, the members participating in the "ba" focused on wisdom for realizing this vision, and in the course of mulling over various ideas, an exchange of knowledge among members was encouraged and new technology and know-how were generated. Had the vision of the Yokohama Smart Community been broader and more abstract, such as "building a smart city" or "realizing a low-carbon society," there is a strong likelihood that the project circumstances would have been quite different. It is because of the very highly specific vision to "establish an energy system learned from a view of life found in nature" that members sharing an affinity with this vision began to participate in the "ba" one after another, and from there exchanges of knowledge took place.

This leads to the other question about the other requirement of sharing the vision among the participating members of the "ba." That is, even if a "ba" with a clear vision is created, if the level of understanding of the vision differs among members participating in the "ba," or if the sharing of that vision is not pervasive in the "ba," the exchange of knowledge among members and the creation of new knowledge will be difficult. The situation that can most likely be imagined in this case is one where principles with a vision that diverges from reality will be adopted, and it will be impossible to obtain the understanding and agreement of the members. In companies, for example, there are many cases where the corporate creed and company motto simply express lofty ideals and flowery words that fail to resonate with the company's employees and effectively become mere expressions with no meaning. Visions adopted by "ba" are the same. Attempting to share a vision among members based only on the promotion of abstract ideals is difficult. What is important is that the vision espoused by the "ba" elicits an affinity among members. As stated earlier, the Yokohama Smart Community promoted a vision with specific, strong message-like content to "establish an energy system learned from a view of life in nature," and continually encouraged companies that shared an affinity with this vision to participate in the consortium. As a result, robust exchanges of knowledge took place and new knowledge was created.

In regard to the nature of the "ba," I also indicated the correlation between the level of freedom of the "ba" and emergence in previous implications. In the case studies in this book, I was also able to confirm the commonly accepted notion and general view that the higher the level of freedom of a "ba," the more apt emergence is to occur. However, a high degree of freedom is not an essential requirement for inducing the exchange of knowledge at "ba." As indicated in the demonstration for the Next-generation Transport Systems in the Yokohama Smart City Project, there are times when robust exchanges of knowledge among members take place even in "ba" where there level of freedom is low. If the vision is firmly shared among the members participating in the "ba," exchanges of knowledge that lead to value creation can occur even in conditions where the level of freedom of the "ba" is restricted. More important than the level of freedom of a "ba" is the management of the "ba."

In other words, management of the "ba" is an issue relating to the leadership of the leader who controls the "ba." Already previous implications indicated the importance of the existence of a "ba" leader but did not discuss the details of that leadership. Let us now consider the nature of the leadership of a "ba" leader that encourages the exchange of knowledge among members in a "ba" that leads to value creation. Research on leadership in the area of management is being actively pursued by scholars and there is a growing body of theoretical studies. However, hierarchical organizations are the subject of the majority of this research, which differ in nature from leadership of a "ba." For example, the "ba" in the smart city projects that are the subject of analyses of this book are flat spaces antithetical to hierarchical organizations, and superior-subordinate relationships among participating members basically do not exist. Therefore, leadership based on downward communication does not apply.

In this regard, the case of the Fujisawa SST offered valuable clues. Through leadership that integrated top-down and bottom-up approaches, Panasonic, the "ba" leader of the Fujisawa SST, promoted the exchange of knowledge among members and indicated a method of "ba" management that leads to value creation. Here I would like to call attention to three points that characterize the essence of that approach. The first is that the leader must be given authority and responsibility even in a flat organization where subordinate-superior relationships are rare. Otherwise the other participating members will not follow the leadership shown by the leader. Since Fujisawa SST was a project initiated by Panasonic, the authority of the leader was clear.

The second point is the leader's indication of a basic direction in value creation. While respecting the participating members' free thinking and the exchange of knowledge based on this without any intervention on the part of the leader is desirable in "ba" with a high level of freedom, if there is a change in viewpoints, there is a risk that members will seek to have their own way and that the situation will go out of control. In the Fujisawa SST, Panasonic indicated to other members the design of the overall project and the values to be created, and the other members engaged in project activities from their respective positions with an understanding of the basic direction as indicated by the leader. This method is an effective form of management in the smooth promotion of cooperation, particularly among companies of different sectors.

The third point is the leader's care in allowing for the energetic exchange of knowledge among members by refraining from interfering as much as possible in the individual activities of the respective members and by utilizing the knowledge the members had. In the Fujisawa SST, the knowledge of the participating members was used to advantage, a "ba" was created where the exchange of knowledge among members promoted, and, even though the final consolidation of the project was the role of Panasonic, freedom in the members' activities was secured at the "ba."

Here I would like to mention the integration of analysis and synthesis as requirements for establishing a "competitive advantage through co-creation." As indicated earlier, co-creation was originally an action expressed through a synthesis approach, and an analytical approach was understood to be the opposite action. In reality, however, it is more appropriate to consider co-creation as occurring when analysis and synthesis merge. In other words, it is believed that co-creation occurs during the integration of the process of analysis where an overall design that has been prepared proceeds in a top-down manner and the process of synthesis where knowledge and various ideas that have arisen in the workplace connect. In the three case studies taken up in this book, the Yokohama Smart City Project can be categorized as an analysis model, the Yokohama Smart Community as a synthesis model, and the Fujisawa SST as an integrated analysis and synthesis model. The case study in which co-creation among companies of different sectors was most notably manifest was the Fujisawa SST. In the case of the Fujisawa SST, although the orientation of the project and the values to be created were clearly defined beforehand and the project proceeded according to an overall design, a "ba" for the exchange and integration of knowledge and know-how among companies was created, and value co-creation took place in an environment where there was an equilibrium between top-down and bottom-up approaches. The question as to at what ratios analysis and synthesis should be integrated to facilitate the most value co-creation is an interesting subject but there is no set mathematical equation that will offer such a solution to this question. The appropriate answer may be that it is a case-by-case situation depending on the particular example. Nevertheless, although there is no single solution, certain logic does exist. In other words, when either analysis or synthesis is too strong, rather than creating a relationship of power where one places pressure on the other, both have only the power to compete with each other.

References

Andrews, K. R. (1971). *The Concept of Corporate Strategy*. Homewood, III: Dow Joes-Irwin.

Ansoff, H. I. (1965). *Corporate Strategy: An Analytical to Business Policy for Growth and Expansion*. New York: McGraw-Hill.

Bower, J. L., & Gilbert, C. G. (Eds.). (2007). *From Resource Allocation to Strategy*. Oxford: Oxford University Press.

Burgelman, R. A. (1983). A process model of internal corporate venturing in the diversified major firm. *Administrative Science Quarterly, 28*(2), 223–244.

Burgelman, R. A. (1994). Fading memories: A process theory of strategic business exit in dynamic environments. *Administrative Science Quarterly, 39*(1), 24–56.

Burgelman, R. A. (2002). Strategy as vector and the inertia of coevolutionary lock-in. *Administrative Science Quarterly, 47*(2), 325–357.

Carley, M., & Christie, I. (1992). *Managing Sustainable Development*. London: Earthscan.

Christensen, C. M., & Raynor, M. (2003). *The Innovator's Solution: Creating and Sustaining Successful Growth*. Boston, MA: Harvard Business School Press.

Dacin, T., Hitt, M., & Levitas, E. (1997). Selecting partners for successful international alliances: examinations of US and Korean firms. *Journal of World Business, 32*(1), 3–16.

Das, T., & Teng, B. (1998). Between trust and control: Developing confidence in partner cooperation in alliances. *Academy of Management Review, 23*(3), 491–512.

Eisenhardt, K. M., & Santos, F. M. (2002). Knowledge-based View: A new theory of strategy. In A. M. Pettigrew, H. Thomas & R. A. Whittington (Eds.). *Handbook of Strategy and Management*. London: Sage.

Friedman, M. (1962). *Capitalism and Freedom*. Chicago: University of Chicago Press.

Gray, B. (1985). Conditions facilitating interorganizational collaboration. *Human Relations, 38*, 911–936.

Itami, H. (1999). *Management of "Ba": A New Management Paradigm*. Tokyo: NTT Publishing.

Kanter, R. (1994). Collaborative advantage: Successful partnerships manage the relationships, not just the deal. *Harvard Business Review, 72*(4), 96–108.

Lave, J., & Wenger, E. (1991). *Situated Learning: Legitimate Peripheral Participation*. Cambridge: Cambridge University Press.

Lazer, D., & Friedman, A. (2007). The network structure of exploration and exploitation. *Administrative Science Quarterly, 52*, 667–694.

Mintzberg, H. (1973). Strategy-making in three modes. *California Management Review, 16*(2), 44–53.

Mintzberg, H. (1978). Patterns in strategy formulation. *Management Science, 24*(9), 934–948.

Mintzberg, H. (1990). The design school: Reconsidering the basic premises of strategic management. *Strategic Management Journal, 11*, 171–195.

Mintzberg, H., & Waters, J. A. (1985). Of strategies. *Deliberate and Emergent, Strategic Management Journal, 6*, 257–272.

Nonaka, I., & Takeuchi, H. (1995). *The Knowledge Creating Company*. New York: Oxford University Press.

Nonaka, I., & Konno, N. (1998). The concept of "Ba": Building a foundation for knowledge creation. *California Management Review, 40*, 40–54.

Nonaka, I., Toyama, R., & Hirata, T. (2008). *Managing Flow: A Process Theory of the Knowledge-Based Firm*. London: Palgrave macmillan.

Porter, M. E. (1980). *Competitive Strategy*. New York: Free Press.

Porter, M. E. (1985). *Competitive Advantage: Creating and Sustaining Superior Performance*. New York: Free Press.

Porter, M. E. (1991). Toward a dynamic theory of strategy. *Strategic Management Journal, 12*, 95–117.

Porter, M. E. (1996). What is strategy? *Harvard Business Review*, *74*(6), 61–78.

Porter, M. E., & Mark, R. K. (2011). Creating shared value: How to reinvent capitalism and unleash a wave of innovation and growth. *Harvard Business Review*, 1–17.

Quinn, J. B. (1978). *Strategic Change: Logical Incrementalism* (pp. 7–21). Fall: Sloan Management Review.

Quinn, J. B. (1980). *Strategic for Change: Logical Incrementalism*. Engelwood Cliffs, NJ: Irwin.

Steiner, G. A. (1969). *Top Management Planning*. New York: Macmillan.

Vangen, S., & Huxham, C. (2003). Nurturing collaborative relations: Building trust in international collaboration. *The Journal of Applied Behavioral Science, 39*(1), 5–31.

Weick, K. E. (1989). Theory construction as disciplined imagination. *Academy of Management Review, 14*(4), 516–531.

Weick, K. E. (1995). *Sencemaking in Organizations*. London: Sage.

Wenger, E. C. (1998). *Community of Practice: Learning, Meaning and Identity*. Cambridge: Cambridge University Press.

Wenger, E. C., & Snyder, W. E. (2000). Communities of practice: The organizational frontier. *Harvard Business Review, 78*(1), 139–145.

Wistow, G., & Hardy, B. (1991). Joint management in community care. *Journal of Management in Medicine, 5*, 40–48.

Chapter 7
Conclusion

In concluding this book, I would like to stress once again the significance of the creation of innovation by the corporate world in establishing a low-carbon society and I would like to summarize issues in establishing competitive advantage through co-creation.

7.1 The Significance of the Establishment of a Low-Carbon Society and the Creation of Innovation by the Corporate World

The need to achieve a low-carbon society has long been identified as a pressing issue but CO_2 emissions of the world as a whole continue to increase. Since 1995, the United Nations has been holding the United Nations Framework Convention on Climate Change (COP) every year and engaged in repeated discussions on the creation of an effective framework for initiatives to address the problem of global warming around the world. Furthermore, the report of the Intergovernmental Panel on Climate Change (IPCC) has issued repeated warnings of the dangers caused by global warming based on scientific data. Despite this, the world remains unable to resolve this problem and the situation is growing increasingly worse.

In view of this situation, what methods would be effective in addressing this problem? It was with a basic awareness of this problem that I began writing this book. The conclusion I arrived at was the creation of innovation by business entities. In other words, I felt that the creation of innovation by the corporate world would be the most effective approach to solving the problem of global warming. If we look at past changes in society, we can find evidence of society making significant transitions and evolving into a new society due to the creation of groundbreaking innovative technology, products and services that have the power to change society. This applies to the invention of electricity, the appearance of the automobile, and the dissemination of IT. Rather than international agreements decided through political bargaining, I believe the driving force of social change is innovation brought about by companies that create innovative technologies,

© Springer Japan 2016

N. Tokoro, *The Smart City and the Co-creation of Value*,
SpringerBriefs in Business, DOI 10.1007/978-4-431-55846-0_7

products and services. Based on this believe, it stands to reason that if the market were saturated with low-carbon products created by companies, such as electric cars and solar-powered products, society should be able to make a significant shift from the current high-carbon society toward a low-carbon society. If this is to happen, we need to consider the kind of environment and conditions that are conducive to the creation of innovation by companies. As the author writing this book, I have had a consistent awareness of this problem from beginning to end.

In the establishment of a low-carbon society, irrespective of whether the innovation companies create is in the form of products or services, the linking of individual products and services with each other will be vital, unlike products and services in the past that were generally created as standalone products. While electric vehicles and solar panels are innovative products on their own, their integrated use with other products rather than their existence as solitary products dramatically increases energy efficiency. For example, if solar panels installed on the roof of a home are connected to an electric car in the garage, electricity generated during the day can be transmitted to the car for use as an energy source or can be stored in a lithium-ion battery for use as electricity in the home in the evening. Endeavors to convert an entire city to low-carbon technology by linking various individual products and services in this way constitute the smart city framework. If cities around the world can be converted into smart cities, we should be able to make significant inroads into solving the problem of rising temperatures.

In such endeavors, it is the innovation created by companies that holds the key to the construction of smart cities. Various products and services based on the key concept of carbon reduction can only be realized through the creation of innovation by companies. Moreover, the "links" between products and services will be achieved not through the creation of innovation by a single company on its own but through the collaborative co-creation of a number of companies hailing from different industrial sectors.

Bearing in mind such a view, I set as the main theme of this book the establishment of competitive advantage through the co-creation of companies of different sectors. That is, the main argument of this book is that while innovation created by companies is essential for today's society to make the transition from the current high-carbon society to a low-carbon society, the establishment of competitive advantage derived from the co-creation of companies is essential for that to occur. If co-creation affords opportunities to companies to acquire competitive advantages, it will become an incentive for companies to actively engage in initiatives to solve this problem, and companies will be the driving force in creating innovation. With such an awareness of this problem, this book closely examined the potential of establishing competitive advantage based on co-creation through the analyses of smart city projects. Many of the smart city projects, however, are still in the early stages and, understandably, there were areas where adequate implications could not be drawn in investigations presented in this book. This was particularly true in regard to the creation of economic value, where validation will require a somewhat longer time axis. I truly hope that in the future many researchers, particularly researchers of management will recognize the importance of the theme taken up in this book and will vigorously pursue theoretical and empirical research on the establishment of competitive advantage through co-creation. Of course, I also intend to continue to devote efforts to research on this theme.

7.2 Issues in Establishing Competitive Advantage Through Co-Creation

Next, I would like to mention problem areas in the theme of establishing competitive advantage through co-creation. One is the extent to which companies are open to disclosing details of their companies' intellectual assets including proprietary technology and know-how as they engage in co-creation with other companies. Many companies understandably harbor a sense of caution about being open with other companies regarding valuable intellectual assets they own. In the smart city projects taken up in this book too, companies engaged in co-creation in technology only in interface areas for linking the products of their respective areas, and they exchanged between them mutually binding agreements regarding the inviolability of the core areas of their technology.

The issue of safeguarding intellectual assets is a significant impediment that obstructs the development of constructive scenarios for promoting co-creation between companies, generating new value through co-creation, and establishing competitive advantage. If a company that strictly guards its intellectual assets out of fear that they may be stolen by other companies, and becomes preoccupied with partial mutual complementarity with other companies, it will fail to achieve full-scale co-creation, and in the end the establishment of any competitive advantage will be limited. This issue has been discussed in the field of strategy as "open and closed" strategy. Essentially, the issue here is determining from a strategic point of view how open and how closed a company should be with other companies in regard to disclosing its proprietary technology and know-how.

This "open and closed" strategy is reminiscent of various strategies Toyota adopted in regard to its hybrid car and fuel cell car. In 1997 Toyota developed the Prius, the world's first hybrid car. At that time it kept under lock and key its hybrid technology that was capable of switching between a gasoline engine and motor through computer control, and promoted a water-tight closed strategy out of fear of the outflow of its technology to rival companies. As a result, the hybrid car failed to gain a broad customer base in the market and its share of the market has remained a modest few percent. Learning from its past experience, Toyota took a completely different approach when it launched the MIRAI on the market as the world's first fuel cell car in 2014. At that time, it adopted a strategy to disclose its technology and know-how concerning the fuel cell to other companies. Its aim in doing so is said to have been to encourage other companies to enter the market by making fuel cell technology and know-how open and thereby broadening the market for fuel cell cars. Nevertheless, this does not mean that Toyota was open about every aspect of technology and know-how it possesses in regard to its fuel cell car. Needless to say, it has not made public those areas that would be better kept as confidential information in terms of profit for the company. In other words, Toyota made a discreet decision after determining which parts of its intellectual assets it would make open and which parts it would maintain as confidential from the perspective of its own profitability and competitive advantage.

Accurately implementing an "open and closed" strategy requires sophisticated strategic thinking. Sophisticated strategic thinking is necessary to determine whether co-creating with other companies has more potential to create greater value or, conversely, whether keeping the company's proprietary knowledge under lock and key has the potential to further enhance the value of the company's proprietary technology and know-how. Therefore, those responsible for implementing a company's strategy must have the capability of being able to accurately judge the value and future potential of the company's technology and know-how. A company's obsessive guarding of its technology and know-how simply out of fear of it being stolen will diminish the potential effects of co-creation. On the other hand, this does not mean that a company should be entirely open with its intellectual assets. To make the establishment of a competitive advantage through co-creation feasible, a company must be able to accurately determine how open or closed it should be in sharing its intellectual assets.

Index

© Springer Japan 2016

N. Tokoro, *The Smart City and the Co-creation of Value*,
SpringerBriefs in Business, DOI 10.1007/978-4-431-55846-0

MIX
Papier aus verantwortungsvollen Quellen
Paper from responsible sources
FSC® C105338

If you have any concerns about our products,
you can contact us on
ProductSafety@springernature.com

In case Publisher is established outside the EU,
the EU authorized representative is:
Springer Nature Customer Service Center GmbH
Europaplatz 3, 69115 Heidelberg, Germany

Printed by Libri Plureos GmbH
in Hamburg, Germany